99-WWF-7

DEVELOPING AND IMPLEMENTING A PERFORMANCE MEASUREMENT SYSTEM: VOLUME II

by

Terrance M. Brueck, P.E.

EMA, Inc.

2005

WERF

Awwa Research Foundation

Advancing the Science of Water®

The Water Environment Research Foundation, a not-for-profit organization, funds and manages water quality research for its subscribers through a diverse public-private partnership between municipal utilities, corporations, academia, industry, and the federal government. WERF subscribers include municipal and regional water and wastewater utilities, industrial corporations, environmental engineering firms, and others that share a commitment to cost-effective water quality solutions. WERF is dedicated to advancing science and technology addressing water quality issues as they impact water resources, the atmosphere, the lands, and quality of life.

For more information, contact:
Water Environment Research Foundation
635 Slaters Lane, Suite 300
Alexandria, VA 22314-1177
Tel: (703) 684-2470
Fax: (703) 299-0742
www.werf.org
werf@werf.org

This report was co-published by the following organizations. For nonsubscriber sales information, contact:

Water Environment Federation
601 Wythe Street
Alexandria, VA 22314-1994
Tel: (800) 666-0206
Tel: (703) 684-2452
Fax: (703) 684-2492
www.wef.org
pubs@wef.org

IWA Publishing
Alliance House, 12 Caxton Street
London SW1H 0QS, United Kingdom
Tel: +44 (0) 20 7654 5500
Fax: +44 (0) 20 7654 5555
www.iwapublishing.com
publications@iwap.co.uk

Library of Congress Catalog Card Number: 2004114166

WEF ISBN: 978-1-57278-229-7 IWAP ISBN: 978-1-84339-732-8

This report was prepared by the organization(s) named below as an account of work sponsored by the Water Environment Research Foundation (WERF), the Awwa Research Foundation (AwwaRF), and the Seattle, WA Public Utilities. Neither WERF, AwwaRF, Seattle Public Utilities, members of those organizations, the organization(s) named below, nor any person acting on their behalf: (a) makes any warranty, express or implied, with respect to the use of any information, apparatus, method, or process disclosed in this report or that such use may not infringe on privately owned rights; or (b) assumes any liabilities with respect to the use of, or for damages resulting from the use of, any information, apparatus, method, or process disclosed in this report.

EMA, Inc.

This document was reviewed by a panel of independent experts selected by WERF. Mention of trade names or commercial products does not constitute WERF endorsement or recommendations for use. Similarly, omission of products or trade names indicates nothing concerning WERF's positions regarding product effectiveness or applicability.

ACKNOWLEDGMENTS

The research team wishes to thank the participating utilities that contributed much time and effort to development of performance measures in their own organizations, especially the City of Phoenix Water Services (AZ), Union Sanitary District (CA), Seattle Public Utilities (WA), and Central Contra Costa Sanitary District (CA), and for their sharing some of the results of their efforts in this report. In addition, many other utilities participated in sharing experiences of their own with performance measurement systems through WEF/WERF conference workshops and a WERF web conference. These contributions all led to the body of knowledge and experiences presented herein.

We wish to acknowledge the guidance and patience provided by WERF and the project subcommittee over the extended duration of this project, especially their support for continued research activity.

Report Preparation

Principal Investigator:
Terrance M. Brueck, P.E.
EMA, Inc.

Project Team:
Claude Williams
Alicia Crumpton
Penny Brink
EMA, Inc.

Project Subcommittee

Tyler Richards, *Chair*
Department of Public Utilities, Gwinnett County, GA

Dean Falkner
Rock River Water Reclamation, IL

Scott Haskins
Seattle Public Utilities, WA

Ed McCormick
East Bay Municipal Utility District, CA

Wiff Peterson
Aquarion Services

Jon Schellpfeffer
Madison Metropolitan Sewerage District, WI

Water Environment Research Foundation Staff

Director of Research: Daniel M. Woltering, Ph.D.
Director of Technical Services: Amit Pramanik, Ph.D.

Awwa Research Foundation

Senior Project Manager: Linda Reekie

Pilot Utility Contribution

City of Phoenix Water Services Department
Ed Blundon, *Assistant Director, Administration*
Bob Hollander, *Compliance and Regulatory Affairs Administrator*
Steve Schebler, Proj*ect Manager/Civil Engineer*

Union Sanitary District
Steve Hayashi, *General Manager*
David Livingston, *Treatment and Disposal Workgroup Manager*
Donna Wies, *Quality Coordinator*

Seattle Public Utilities
Scott Haskins, *Deputy Director for Field Operations*
Diann Shope, *Strategic Advisor*
Mike Brown, *Management Systems Analyst*

Central Contra Costa Sanitary District
Jim Kelly, *Director of Plant Operations*

ABSTRACT AND BENEFITS

Abstract:

The primary purpose of this research was to provide methods and tools that enable a water/wastewater utility to develop and implement a performance measurement system based on a demonstrated, proven approach. To achieve the purpose, the research was conducted in three phases:

- Core research on existing and applicable frameworks for performance measurement with experience from utilities and other government organizations, as well as outside business and industry, on leading practices in performance measurement. Steps to develop and implement performance measurement that uniquely fit water/wastewater utilities were defined to be piloted by selected utilities in Phase II.
- Demonstration pilot projects for developing and implementing performance measurement were carried out over a 12–18 month period at four water and wastewater utilities. These four utilities were selected, based on a mix of utility type, size, and experience with performance measurement, from a group of more than a dozen that applied to conduct the pilot demonstrations.
- Research results from the pilots and lessons learned were applied to adjust the piloted process. The knowledge and methods from these projects were shared in WERF/WEFTEC workshops and a Web conference.

A process (seven-step methodology) was developed based on a balanced scorecard (BSC) approach to develop and implement performance measures both at the enterprise (utility-wide) level and team-based level. Ways to align and coordinate measures throughout the organization were defined for process-based and initiative-based scorecards. The testing of this approach through the utility pilots led to further recommendations for involvement, education, communication, and commitment of utility participants for successful performance measurement.

Benefits:

- Provides methods and tools for water/wastewater utilities to adapt an approach for performance measurement that fits their organization and current business processes and initiatives.
- Creates a balanced system of measures both at the top and on the ground that relate and align for utility-wide performance improvement.
- Shows examples of how to implement measures using various approaches with samples of measurement scorecards.
- Gives guidance and lessons learned on organizational and cultural aspects of performance measurement.
- Describes technology to support scorecard development and performance measurement tracking, reporting, and automating.

Keywords: Performance measurement, performance management, balanced scorecard

TABLE OF CONTENTS

LIST OF TABLES

LIST OF FIGURES

LIST OF ACRONYMS

APQC	American Productivity and Quality Center
AWWA	American Water Works Association
AwwaRF	American Water Works Association Research Foundation
BSC	Balanced Scorecard
CCCSD	Central Contra Costa Sanitary District
CIS	Customer Information System
CMMS	Computerized Maintenance Management System
EAI	Enterprise Application Integration Tools
EIS	Data Warehouses/Enterprise Information Systems
ERP	Enterprise Resource Planning
FIS	Financial Information System
GASB	Government Accounting Standards Board
GIS	Geographical Information System
HRIS	Human Resource Information System
LIMS	Laboratory Information Management System
LOB	Line of Business
MG	Million Gallons
MGT	Management Leadership Team
ML	Million Liters
PDCA	Plan-Do-Check-Act
POS	Plant Operations Scorecard
SCADA	Supervisory Control and Data Acquisition
SPU	Seattle Public Utilities
T&D	Treatment and Disposal
UBA	Utility Business Architecture
USD	Union Sanitary District
WMS	Work Management System or Maintenance Management System
WSD	Water Services Department

EXECUTIVE SUMMARY

ES.1 Purpose and Approach

The primary purpose of this research was to provide methods and tools that enable a water or wastewater utility to develop and implement a performance measurement system based on a demonstrated, proven approach. To achieve the purpose, the research was conducted in three phases:

- Core research on existing and applicable frameworks for performance measurement including experience and leading practices used within utilities and other government organizations, as well as business and industry. Steps to develop and implement performance measurement that uniquely fit water/wastewater utilities were defined to be piloted by selected utilities in Phase II.
- Demonstration pilot projects for developing and implementing performance measurement were carried out over a 12–18 month period at four water and wastewater utilities. They were selected from a group of over a dozen utilities based on a mix of utility type, size, and experience with performance measurement.
- Research results from the pilots and lessons learned were applied to adjust the piloted process. The knowledge and methods from these projects were shared through WERF/WEFTEC workshops and a web conference.

ES.2 Need for Performance Measurement System

Well-designed and properly implemented performance measurement systems can enable utilities to achieve new levels of performance in terms of efficiency, quality, and effectiveness. The drivers for change can be external—such as competition, regulatory demands, or customer expectations—or can be internally motivated based on recognized opportunities for improvement. Whether internally or externally driven, performance improvement requires measurement. Interest in performance measurement is increasing in businesses and industries today and has been advanced through concepts such as the balanced scorecard (BSC) (Kaplan & Norton, 1996). Utilities can employ these same concepts and learn best practices from other utilities as well as from industries' experiences.

A successful performance measurement system combines a holistic approach around improved business practices and effective human/organizational strategies in addition to performance measures for operational decision-making. Effective performance measurement enables an organization to establish a baseline for its current practices, set targets for change, and drive initial and continuous incremental improvements.

As a basis for continuous improvement, performance measurement supports and is a key part of many business management systems. The continuous improvement cycle that is world-renowned from the work of W. Edwards Deming, as applied in Japan starting in the 1950s, includes measurement as a key element. As Deming's (1993) infamous "red bead experiment" has shown to measurement practitioners worldwide, an understanding of measurements' relationships to process is fundamental to improvement. The plan-do-check-act (PDCA) cycle

(Shewhart cycle) that Deming applied in many industries as the basis for continuous improvement, is rooted in performance measurement. The performance measurement system is a basic tool to use together with the PDCA cycle.

The purpose of a performance measurement system is for utilities to continuously (as well as initially) improve their business performance. Often, the term 'system' is used to denote the information technology that automates data collection, data manipulation, and the recording and presenting of measurement results. For a performance measurement system to be most successful, the system must also include processes, procedures, communication, education, and supporting tools. All of the steps from defining a utility's strategic plan to day-to-day use of performance measures are part of the system.

As shown in Figure ES-1, the components of a performance measurement system are centered on the business process and associated work practices that deliver the product or service to its customers.

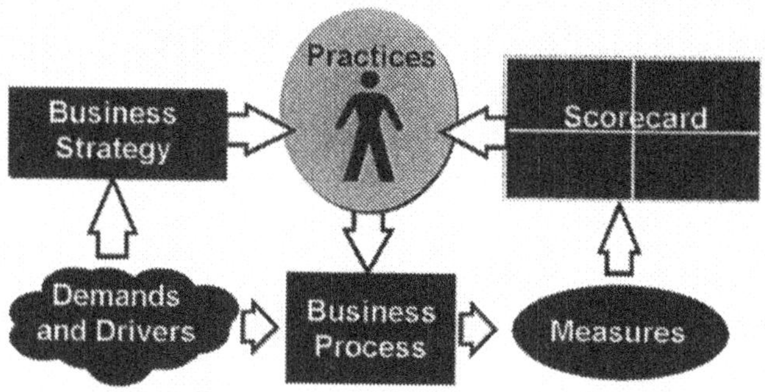

Figure ES-1. A Successful Performance Measurement System Is Based on Well-Understood Business Processes and Associated Work Practices.

ES.3 Performance Measurement Framework

One fundamental concept in understanding performance measurement is putting various measurements into perspective by using a performance measurement framework. A good framework identifies measures that tell to what extent the organization's vision and strategy are being achieved.

For this project, the BSC was selected and modified to be the basis for a performance measurement framework for water/wastewater utilities. The BSC framework is based on the understanding that an organization's continuing success relies on more than financial factors (Figure ES-2). The other three factors include customer, internal business process, and learning and growth. The BSC performance measures from four perspectives have been developed to reflect a balanced view of a public service organization.

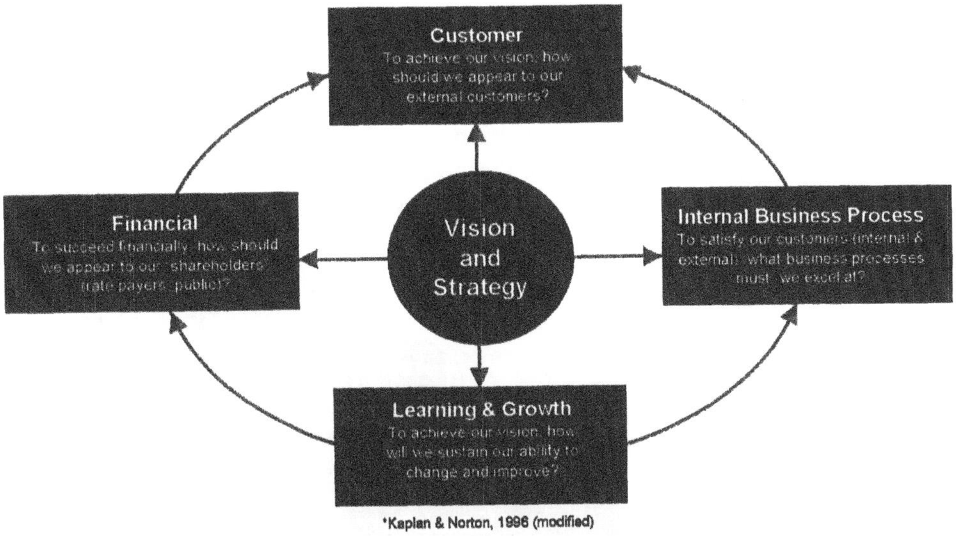

Figure ES-2. Most Widely Used Framework: Balanced Scorecard Was Used.

Another fundamental concept of performance measurement is selecting measures of the right type. Some measures record the input to what was done (cost, effort, number of work orders, or other activity), while other measures indicate results (amount of wastewater treated per dollars spent or percent of wastewater treated meeting effluent quality standards). A good measure should be easy to understand and hard to falsify.

A scorecard should be balanced on measures of efficiency, quality, and effectiveness at each level of the performance measurement system, whether at the enterprise level or at a process or initiative level. The types of measures include efficiency, quality, and effectiveness (Figure ES-3).

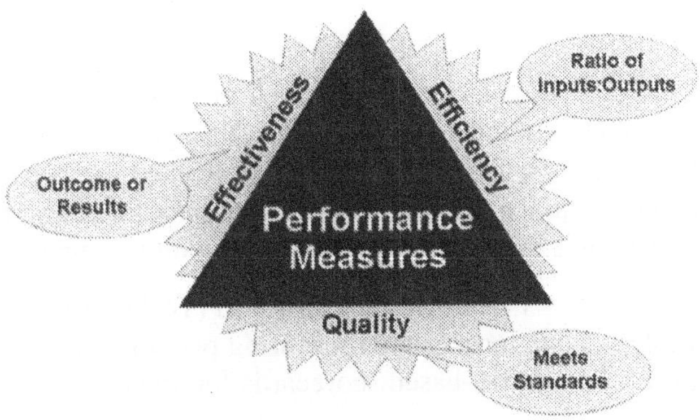

Figure ES-3. Good Performance Measurement Applies Three Types of Measures within Each Scorecard Perspective.

ES.4 Steps to Develop and Implement

A seven-step process for developing and implementing a performance measurement system was created from best practice information and other secondary research completed during Phase 1 of the project. Completion of these seven steps is essential to successfully developing a performance measurement system that results in improved performance (Figure ES-4).

Figure ES-4. Seven-Step Process Provides Common Approach for Developing and Implementing a Performance Measurement System.

ES.5 Demonstration Pilot Projects

In Phase II, four utilities piloted a seven-step process to develop performance measurement for their own unique circumstances. The pilot project utilities and their individual focus areas were:

- City of Phoenix Water Services Department—A utility-wide scorecard was developed and linked to an updated department strategic plan. Implementation of scorecard goals, objectives, and measures was performed for two focus areas: managing maintenance at Union Hills Water Treatment Plant and prevention of sanitary sewer overflows for the zero-tolerance regulatory compliance program.
- Union Sanitary District—A treatment and disposal (T&D) process scorecard focused work teams on establishing targets and measuring performance. A refined scorecard was linked to process and team-based scorecards for an enterprise view of utility-wide performance.
- Seattle Public Utilities—An updated strategic business plan directly linked measures from operational work plans and strategic initiatives to strategic business plan goals. Refinement of the existing organizational performance system focused on balancing the types of measures and implementing lines-of-business measures.

- ♦ Central Contra Costa Sanitary District—An initial plant operations scorecard was defined with identified key performance improvement areas. Additional development and implementation of the scorecard aligned with the District's updated strategic direction.

ES.6 Measuring Performance Utility-Wide

To create a system of utility-wide measures that are related and linked to a utility's goals and objectives requires starting with the vision and mission. As shown in Figure ES-5, the business strategies that support the utility's strategic direction can be tracked by an enterprise scorecard of performance measures.

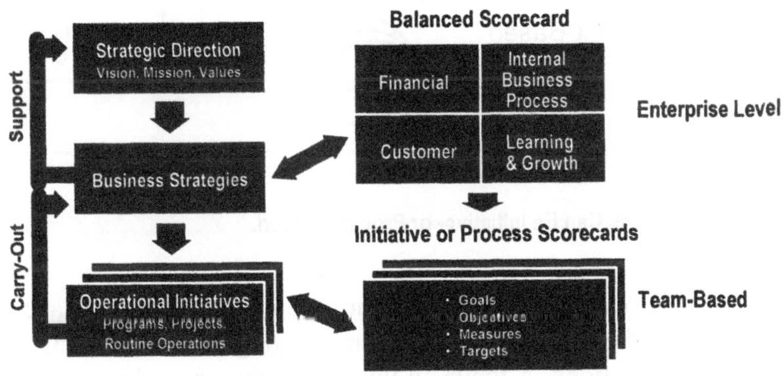

Figure ES-5. Scorecards Can Measure and Relate Performance at the Enterprise as Well as Team-Based Levels.

In addition, each of a utility's operational initiatives that carry out the business strategies can be measured by team-based (process or initiative) scorecards. Processed-based scorecards are linked to a business process framework that defines typical water/wastewater utility functions such as the utility business architecture (UBA). Initiative-based scorecards are linked to an organization's key or strategic initiatives or programs, typically through a utility's strategic planning process. This utility-wide system of measures with related scorecards can be used to apply performance measurement on an enterprise basis, with meaningful team-based measures that link to the utility's goals and objectives. The challenge of relating and integrating this combined approach of enterprise and team-based levels can be addressed in many ways. Several of these approaches were applied to the utility pilot projects and are described in this report.

A utility needs to select the best approach, i.e., initiative-based, process-based, or a hybrid approach to rolling out the scorecards organization-wide as shown in Figure ES-6. Individuals or organization units should identify and have responsibility for a set of initiative-based or process-based measures that may come from multiple scorecards. This handful of individual or organization unit measures must be clearly relevant to their work as well as aligned with achievement of goals and objectives on the scorecards.

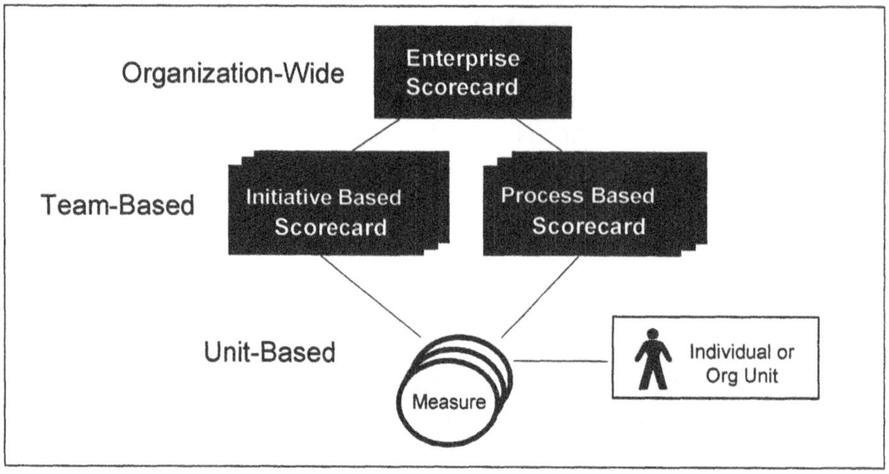

Figure ES-6. Balanced Scorecards Can Be Initiative- or Process-Based.

The seven-step process for developing and implementing performance measurement systems was refined based on real-world experiences using team-based implementations with several scorecard rollout options. A focus on linking on-the-ground measures with enterprise or at-the-top measures led to methods for aligning multiple scorecards with company-wide vision and strategic direction. The research resulted in a flexible implementation process for performance measurement. The process can be adapted based on a utility's current business strategy and need for measures to drive improvement initiatives or continuous process improvements.

CHAPTER 1.0

INTRODUCTION

1.1 Performance Measurement Overview

This joint WERF/American Water Works Association Research Foundation (AwwaRF) research project (Developing and Implementing a Performance Measurement System) has been underway since late 1999. The purpose was to build and test methods and tools that enable a utility to develop and implement a performance measurement system based on a demonstrated, proven approach. An initial report (Volume I) was published in 2000 (WERF). Since then, several utilities were selected to develop and implement performance measurement systems or improve existing systems based on the initial research. This final report recommends an approach for implementing performance measures that can be applied at any utility and summarizes the results of the utility pilot projects.

Well-designed and properly implemented performance measurement systems can enable utilities to achieve new levels of performance in terms of efficiency, quality, and effectiveness. As shown in Figure 1-1, the drivers for change can be external such as competition, regulatory demands, or customer expectations or can be internally motivated based on recognized opportunities for improvement. Whether internally or externally driven, performance improvement requires measurement. Interest in performance measurement is increasing in businesses and industries today and has been advanced through concepts such as the balanced scorecard (BSC) (Kaplan & Norton, 1996). Utilities can employ these same concepts and learn best practices from other utilities as well as industries' experiences.

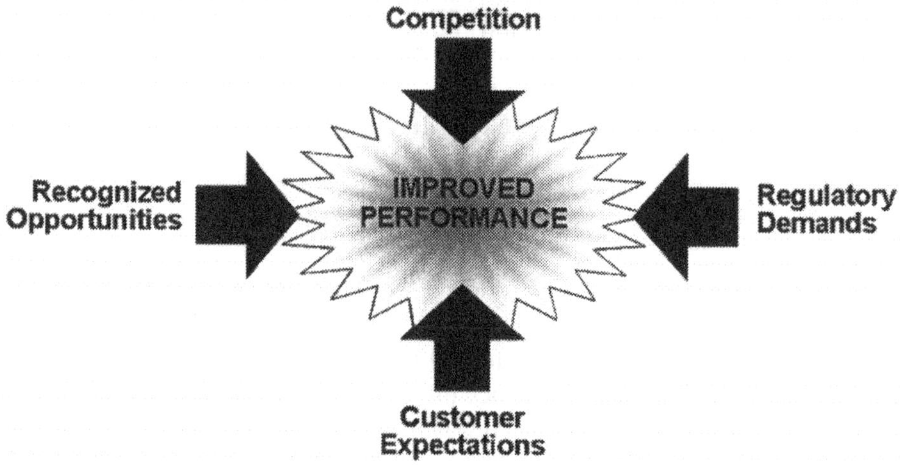

Figure 1-1. The Need to Improve Performance Can Be Driven by Many Factors.

(Note: The term "best practice" has been used extensively in corporate America usually when referring to a known or recognized leading practice. That practice may or may not be best when applied in a specific organization. These practices must usually be adapted and adjusted to the unique needs of an organization to become best practice for that organization. Throughout this document, the terms "best practice" and "leading practice" may be used interchangeably and generally refer to the concept of learning from other leading organizations' experiences).

Performance measurement can be an effective tool in driving organization improvement. A performance measurement system alone does not improve performance or make a utility competitive, but when combined with an appropriate business strategy and performance improvement initiatives, it can show the results of closing the performance gap (Figure 1-2).

The key is to know where your performance is today and set targets for improvement to close the gap. The gap between the two becomes the recognized improvement opportunity. A performance measurement system incorporates the process of defining change drivers, developing goals, objectives, measures and targets, plus tracking those measures over time with actions to improve results.

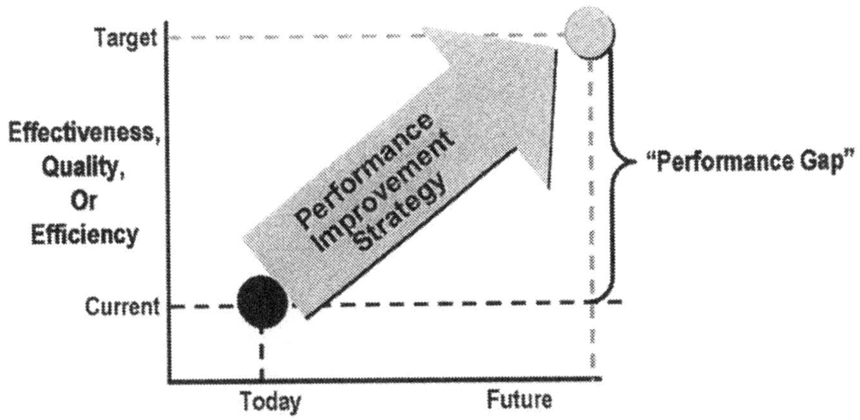

Figure 1-2. Performance Measurement Is Required to Show How Improvement Strategies Close the Performance Gap.

A successful performance measurement system combines a holistic approach around improved business practices and effective human/organizational strategies in addition to performance measures for operational decision making. Effective performance measurement enables an organization to establish a baseline for its current practices, set targets for change, and drive initial and continuous incremental improvements.

As a basis for continuous improvement, performance measurement can support and is key to many business management systems. The continuous improvement cycle that is world-renowned from the work of W. Edwards Deming, as applied in Japan starting in the 1950s, includes measurement as a key element. As Deming's (1993) infamous "red bead experiment" has shown to measurement practitioners worldwide, an understanding of measurement's relationships to process is fundamental to improvement. The plan-do-check-act (PDCA) cycle (Shewhart cycle) that Deming applied (Figure 1-3) in many industries as the basis for continuous improvement is rooted in performance measurement. The performance measurement system as defined herein is a basic tool to use with the PDCA cycle.

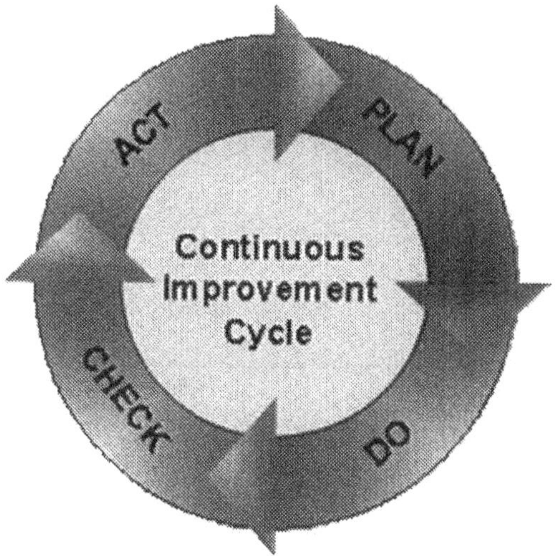

Figure 1.3. The Deming (Shewhart Cycle) for Continuous Improvement Is Based on Performance Measurement.

The purpose of a performance measurement system is for utilities to continuously (as well as initially) improve their business performance. Often, the term *system* is used to denote the information technology that automates data collection, data manipulation, and the recording and presentation of measurement results. For a performance measurement system to be most successful, the system must also include processes, procedures, communication, education, and supporting tools. All of the steps from defining a utility's strategic plan to day-to-day use of performance measures are part of the system.

As shown in Figure 1-4, the components of a performance measurement system are centered on the business process and associated work practices that deliver the product or service to its intended customers (either internal or external customers). These practices and the business process are influenced directly and indirectly by the business environment—the demands and drivers (previously described) that lead to creating business strategies. These business strategies are best if they are explicitly developed based on demands and drivers but may also be based on implicit ways of doing business. For example, a strategy to improve customer service may be driven by external customer demands (e.g., complaints, expectations, reactions) or could be based on internal recognition of the opportunity and ability to improve customer service.

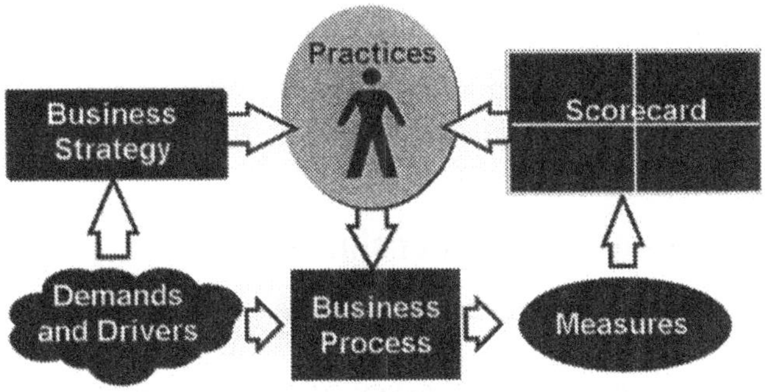

Figure 1-4. A Successful Performance Measurement System Is Based on Well-Understood Business Processes and Associated Work Practices.

As shown in Figure 1-4, to understand the results of these practices and the business process itself requires process-based performance measures in a feedback loop. A feedback tool to understand these measures in terms of expected outcomes or targets is a scorecard. This scorecard (also sometimes referred to as a performance dashboard) is used as feedback to make adjustments in the work practices or business process. A BSC will be further described later in this report. The BSC was selected for the measurement framework in the initial project work (Volume 1 report).

With a system of appropriate measures defined and these measures linked to organizational accountabilities, adjustments can be made to business processes or practices. An ongoing cycle of tracking measures and making process adjustments is the heart of implementing a performance measurement system.

As previously described, the measurement system alone will not necessarily result in improved performance, but with related understanding of business processes and practices and organizational actions, a utility can carry out its performance-improvement strategies. This report will further describe these measurement concepts and framework with a multi-step implementation process. Highlights of other common elements of well-developed systems include:

- Senior management demonstrates routine, visible commitment to the program. Strong leadership creates a positive climate for performance improvements including fostering cross-functional, mission-oriented performance improvements.
- Measurement results are analyzed and published regularly (e.g., minimum of once per month), with performance tracked over time.
- Employees throughout the organization understand specifically how their jobs relate to their business unit performance, which in turn relates to the organization's performance. They are actively engaged in analyzing and using the data provided by the performance measurement system.
- Measures are linked to other management systems, including overarching strategic business planning, operational planning and goal setting, budgeting, and team-based and/or individual incentive programs.

- Measures and targets are revised to reflect changes to the strategic plan, vision, and/or mission, as well as continuous improvement through changes at the team level.
- An ongoing education and communication program is in place to enable employees' understanding of the measures and involvement in business improvements.

1.2 Performance Measurement and Benchmarking

The water and wastewater utility industry has given much attention to benchmarking of performance measures through a number of studies over the last five-to-ten years. Compared to other industries, this is a relatively short and recent emphasis. A number of these studies are highlighted in chronological order in Table 1-1 with references in the bibliography.

Table 1-1. Summary of Significant Benchmarking Efforts in the Water and Wastewater Utility Industry.

Year	Effort
1996	Performance Benchmarking for Water Utilities (AwwaRF, 1996)
	Western Regional Water Utilities Benchmarking Group (Paralez, 1999)
1997	Benchmarking Wastewater Operations (WERF, 1999)
1998	Multi-Agency Benchmarking (West Coast Wastewater Utilities)
1999	AMSA/WERF Cleanwater Central (Initial Release with AMSA Financial Survey Data)
2000	IWA Performance Indicators (IWA, 2000)
2001	QualServe Benchmarking Clearinghouse Performance Indicators (AwwaRF, 2002)
2002	QualServe Pilot Survey on Performance Indicators (QualServe, 2002)
2003	QualServe Performance Indicators Benchmarking Report (QualServe/APQC, 2003)
	CleanWater Central Survey Database (AMSA/WERF, 2003)

While each of these efforts has added to and advanced the body of knowledge on performance measurement for water and wastewater utilities, most of these studies have focused on metric benchmarking. Process and practice benchmarking have received less emphasis and may therefore be less well-understood. A summary of each of these types of benchmarking was included in a report by Brueck, Riddle, and Paralez (AwwaRF, 2001) with a portion excerpted and updated here.[1] This benchmarking guide has been adopted by the AWWA/WEF QualServe Benchmarking Clearinghouse.

> Benchmarking is designed to yield great benefits in the education of managers and executives to apply performance improvements for business operations. Benchmarking can take several forms. First, there is the focus of either "internal" or "external" to the organization. Internal benchmarking studies the practices and performance within an organization. External benchmarking determines the performance of other, preferably world-class, companies.
>
> Second, there is the selection of the type of approach to be used to undertake the specific study. Benchmarking typically consists of separate methods such as "metric

[1] *Consortium Benchmarking Methodology Guide*. Copyright 2003 by AwwaRF. Reprinted with permission.

benchmarking," "process benchmarking," and "practice benchmarking." They are briefly described below and can be undertaken either as a package or independently.

Metric Benchmarking

Metric benchmarking is a quantitative comparative assessment that enables utilities to track internal performance over time and to compare this performance against their past performance baseline or compared to similar utilities. Areas of relatively low performance, compared to that of other utilities, can be identified where there is particular room for improvement in performance. In addition, through the comparison process, target levels of performance can be established.

Current metric benchmarking in the water industry tends to focus on the use of simple performance ratios of inputs to outputs such as cost per mile of pipeline or cost per gallon treated. The diversity of topography, water sources, customer and workforce demographics, ownership and organizational structures, facility and infrastructure conditions, and environmental and financial regulations in the U.S. water industry, invalidates most attempts at comparison of utility performance using simple ratios that do not explicitly account for these "explanatory factors." While ratios may illuminate trends that can be tracked over time, they are often poor measures for inter-utility comparisons.

Ratios can be very reliable indicators of performance over time within a given utility when they are well defined, applied consistently, and set in the proper context. In addition, when metric measurements are made as an integral part of the business practice or process, the resulting metric data becomes useful because it is consistent over time and between the same practices or processes of different organizations. The challenge for the water industry is to develop meaningful "standardized" business practices or processes and associated metrics so that the resulting data becomes comparable between organizations. A significant effort to develop a set of high-level measures for the water and wastewater industry was conducted by the AWWA/WEF QualServe Benchmarking Clearinghouse (AwwaRF, 2003). The initial set of performance measures included twenty-two measures that have been the basis of data collected from a pilot group of 24 utilities in 2002.

Process Benchmarking

Process benchmarking involves first identifying specific work procedures to be improved through a step-by-step "process mapping," and then locating external examples of excellence in these process elements for standard setting and possible emulation. This is also known as "Xerox"-style benchmarking after work undertaken by the Xerox Corporation beginning in 1979. Process benchmarking, unlike metric benchmarking, provides a tool by which utilities can change the way they work by introducing improvements in efficiency and service by comparing and contrasting processes.

Process benchmarking is generally higher-level and less numbers-intensive than metric benchmarking. Process benchmarking studies demonstrate how top performing companies accomplish the specific process in question. Such studies can take the form of research, surveys/interviews, and site visits. By identifying how others perform the same functional task or objective, organizations gain insight and ideas they may not otherwise achieve. Such information affirms and supports quality decision making by executives. This insight is one key benefit and value-added feature of benchmarking.

There are significant opportunities for process benchmarking in water and wastewater utilities where processes are well defined and similar utility-to-utility. An example of some areas where process benchmarking has the potential to yield significant improvements are lab operations, pump repair, pipeline replacement, bill collection, etc. These and similar water and wastewater utility activities have well defined, repeatable, and specialized business processes.

Practices Benchmarking

Business practices benchmarking focuses on gathering profiles of practices and correlative metrics. Such profiles provide a means for organizations to understand what practices—from a range of expected "best" or promising practices—are being used. Companion metrics provide some indication of the results each organization is getting from the implementation of those practices. This was the method used for the AwwaRF Energy Management Practices study (AwwaRF, 2002). This method was similar to most of the consortium studies conducted to-date by the Western Regional Water Utility Benchmarking Group (Paralez, 1999).

Business practices benchmarking is the process of seeking out and studying the best business practices that produce superior performance. The traditional metrics-focused approach is supplemented with an analysis of what or which business practices produce exceptional results. This practices-based approach also reduces some of the effort required for process benchmarking. It avoids the need for some level of process mapping before the benchmarking can be done. However, practices benchmarking requires subject matter expertise to identify the known or potential best practices in advance, so that survey questions can be developed based on those best practices.

The outcomes of practice benchmarking typically are histogram profiles that reveal the extent to which participants use, engage, or benefit from a given set of business practices. When metrics are associated with these business practices, the differences in metrics can often show a utility's performance level relative to the practice. Again, a challenge is to "standardize" the business practices and associated metrics well enough so the results are truly comparable between utilities.

Using survey techniques with a consortium of participants, the resulting metrics or findings of the practices can be used to select case studies of potential best practices. From the case studies, details of the practices, associated business processes, and additional information can be gathered to be shared with consortium participants.

1.3 Benchmarking and the Need to Measure

While each kind of benchmarking may be used alone or in combination for different purposes, they all have in common the need for measures. These measures can be metrics at a high-level or very detailed level and may be different types of measures such as effectiveness, quality, or efficiency. These measures may also be used to qualify a utility's practices compared to leading practices such as measures of frequency or formality of practices in use. No matter what kind of benchmarking is employed, it can jumpstart, augment, or serve as a basis for continual comparison either internally or externally. The measures included in a utility's performance measurement system can be this basis for comparison.

1.4 Why a Performance Measurement System?

A sometimes-common belief in many utilities is that a right or standard set of measures should exist or could be developed to allow for inter-utility comparisons. While inter-utility comparative measures can be helpful for some purposes or in certain circumstances, many times they can be misleading. Among utilities, metric data comparisons alone, as shown in Figure 1-5, fail to address the differences in utilities that explain the differences in performance measures.

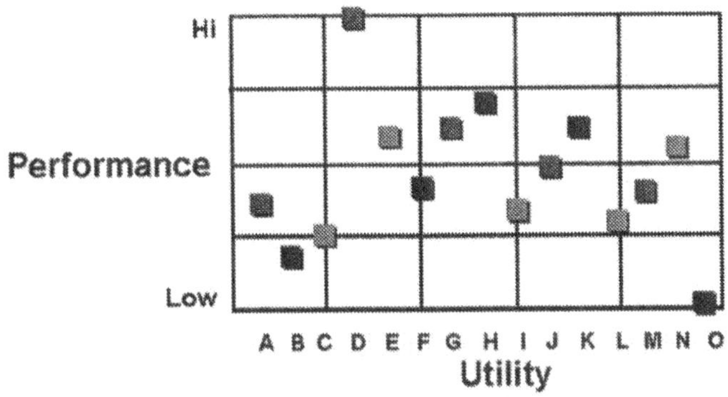

Figure 1-5. Typical Industry Statistics Are Not Standardized or Normalized for Apples-to-Apples Comparison.

These differences or explanatory factors include utility circumstances such as:

- Types of services provided
- Levels of service
- Population served (density, growth, demographics)
- Climate and geography
- Water sources and receiving waters
- Infrastructure and financial conditions
- Labor and energy costs
- Residential/commercial/industrial mix

Attempts to normalize this kind of data, or even be sure the data represents measures that are well enough defined to be comparable, have met with limited success. Complex normalization schemes using statistical methods have failed to produce truly comparable performance measures. Sometimes utilities faced with metric performance data that show their utility as an under performer resort to discrediting the source or collection of the data or simply trying to use explanatory factors to explain why. Conversely, high performers have no need or reason to discredit or explain. Using metrics alone can be misleading; at best, it shows differences, but does not explain the reason for the differences.

As an example, two utilities' metric data are shown in Figure 1-6. In comparing metrics alone, Utility A has the higher productivity in terms of unit output per unit input (e.g., million gallons per day (MGD) produced per daily cost to produce). However, in view of other factors, such as quality of source/influent, standards or regulations for quality of output/effluent, cost of labor, energy, chemicals, and so forth, this purely metric comparison may be insufficient to tell the whole story. While additional, more detailed metric comparisons (e.g., kW/MG, labor $/MG) may reveal differences in performance, it is difficult to uncover the real explanatory factors using only these metric comparisons without knowledge of business processes or work practices.

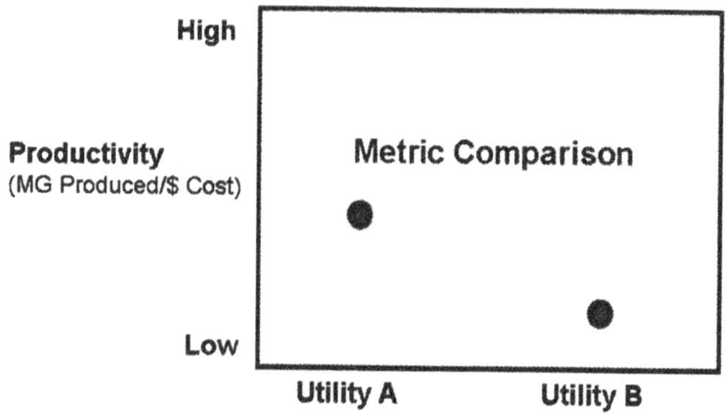

Figure 1-6. Metrics Only Tell Part of the Story in Terms of Performance Comparisons.

To tell the real story in terms of which utility has higher performance, one must consider the factors that are truly controllable versus those that are given. So each utility, given its own set of circumstances (influent quality, effluent standards, cost of commodities, etc.), must assess its use of best practices to deal with these circumstances. In comparing each utility's unique situation and practices in-use, one can determine the gap between current practices and best practices as shown in Figure 1-7.

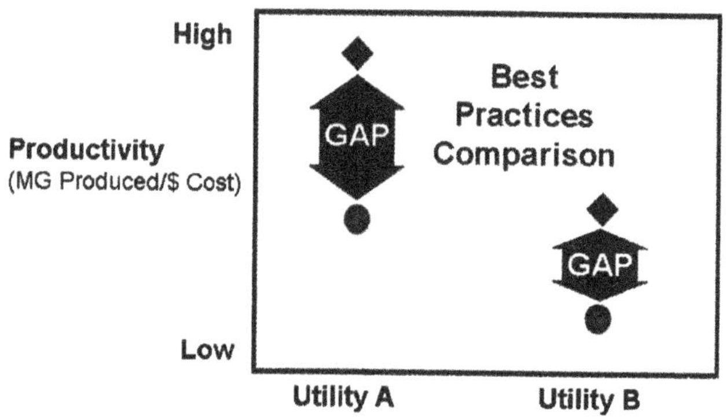

Figure 1-7. Practice-Based Benchmarking Compares Each Utility to Best Practices.

Based on the utility's methods, tools, processes, procedures, resources, etc., a gap can be calculated between current practices and best practices (or other improved/better practices, as best is a moving target). Using this gap comparison between the two utilities because Utility B has the smaller gap based on best practices, it is in fact the more productive or competitive utility.

The real point of this discussion of metric-based comparisons and practice-based comparisons is that a utility needs an ongoing system of performance measures in order to build in continuous improvement.

1.5 WERF/AwwaRF Project Background

As shown in Figure 1-8, this project included three distinct phases of work. Beginning with the core research of Phase I; piloting of performance measurement systems with utility demonstrations in Phase II; and subsequent adjusting of methods based on pilot results with additional knowledge transfer in Phase III, this project has spanned several years of effort.

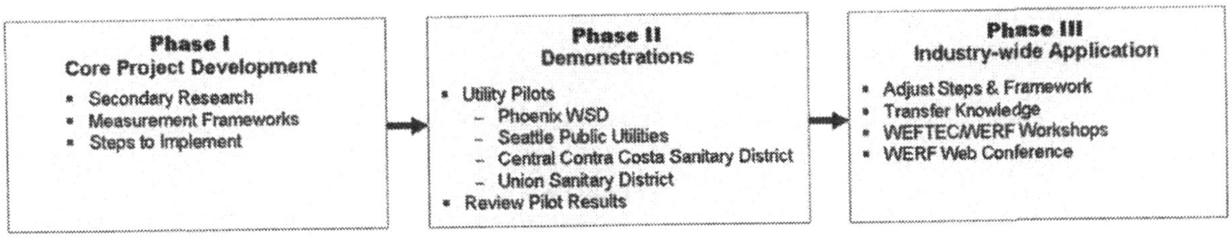

Figure 1-8. WERF/AwwaRF Project "Developing and Implementing a Performance Measurement System" Included Three Phases.

Phase 1: Core Research

Phase I included research on existing and applicable frameworks for performance measurement such as BSC and other frameworks by the Government Accounting Standards Board (GASB), Malcolm Baldridge National Quality Award, and the National Performance Review. Experiences from public and private utilities, other government organizations, as well as outside industry and business experiences were captured for determining performance measurement leading practices (APQC, 1996; 1997a,b; 1999; 2002). Steps to implement performance measurement systems were defined based on this research of best practices. These steps became the basis for Phase II piloting of performance measurement systems in selected utilities.

Based on an application process and set of selection criteria, three utilities were selected to pilot the process for developing and implementing a performance measurement system. A fourth utility was later added as a part of WERF's Targeted Collaborative Research Program.

Phase II: Demonstrations—Pilot Projects

Four utilities piloted a seven-step process to develop performance measurement for their own unique circumstances. Summaries of several of the pilot implementations, including individual utility's lessons learned and adjustments to the performance measurement framework and process, are included as Appendix C and Chapter 4.0. The pilot project utilities and their individual focus areas were:

- City of Phoenix Water Services Department—A utility-wide scorecard was developed and linked to an updated department strategic plan. Implementation of scorecard goals, objectives, and measures was performed for two focus areas: managing maintenance at Union Hills Water Treatment Plant and prevention of sanitary sewer overflows for the zero-tolerance regulatory compliance program.

- Union Sanitary District—A treatment and disposal (T&D) process scorecard focused work teams on establishing targets and measuring performance. A refined scorecard was linked to process and team-based scorecards for an enterprise view of utility-wide performance.

- Seattle Public Utilities—An updated strategic business plan directly linked measures from operational work plans and strategic initiatives to strategic business plan goals. Refinement of the existing organizational performance system focused on balancing the types of measures and implementing lines-of-business measures.

- Central Contra Costa Sanitary District—An initial plant operations scorecard was defined with identified key performance improvement areas. Additional development and implementation of the scorecard aligned it with the District's updated strategic direction.

Two of the pilot utilities, Seattle Public Utilities and Union Sanitary District, have previously implemented organization-wide performance measurement systems. All pilot participants have previously used performance measures in some areas. This project demonstrated an approach that helps utilities begin to embrace or expand performance measurement, regardless of their previous level or use of performance measures.

Phase III: Application—Research Results

Following a demonstration period of between 12 and 18 months at each of the pilot utilities, results of implementing performance measurement and lessons learned were used to adjust the piloted process. A variety of utility types, sizes, and previous experience with performance measures led to a process that can be adapted for widespread application. Summaries of utility experiences were shared in WEFTEC/WERF Conference Workshops during 2001 and 2002 and at a WERF Web Conference in 2003.

The seven-step process for developing and implementing performance measurement systems was refined based on real-world experiences using team-based implementations with several scorecard rollout options. A focus on linking on-the-ground measures with enterprise (at-the-top) measures led to methods for aligning multiple scorecards with company-wide vision and strategic direction. The research resulted in a flexible process for performance measurement implementation. The process can be adapted based on the utility's current business strategy and need for measures to drive improvement initiatives or continuous process improvements.

Additional research would be beneficial in several areas related to performance measurement, including supporting and integrative technologies, ties to incentive systems, and relationships to industry-standard performance indicators. Additional technology transfer and education, including simple tools for utilities to get started with performance measurement, would also add value for the water and wastewater industry.

CHAPTER 2.0

OVERVIEW AND CONCEPTS

2.1 Best Practices Overview

This chapter presents an overview of practices used in measuring and managing performance and explains some concepts that are fundamental to understanding and applying measurement practices. This project investigated a number of best practices from other industries and adapted these practices for water and wastewater utilities for validation with utility pilot projects. From secondary research conducted in other industries, including *Fortune* 500 companies that were included as part of consortium studies (APQC 1999; 1997a,b; 1996; 2002), five key areas were identified for successful performance measurement. Examples of these best practices developed for application with wastewater and water utilities are shown for each area below.

1. Performance measures must be linked, aligned, and integrated with an organization's strategic plan and business initiatives.
   - Performance measurement has strong, clear alignment with an organization's strategic planning process. Reporting of measures reinforces the strategic plan and helps communicate it to all employees.
   - Performance measurement is integrated with quality initiatives (ISO 9000, 14000), award programs (e.g., the Baldridge Award), or performance improvement programs/strategy. The quality initiatives are a specific part of the strategic plan. Performance measurement systems provide a major part of the information required in these initiatives.
   - Visible leadership through senior-level champions. Senior management makes performance measurement an important part of the way business is done and communicates the targets and results on a regular basis.
   - Performance measurement ties to incentive programs, pay for performance, or at-risk compensation, especially at the executive and senior levels. The performance measures are often for a team or business unit but may also include personal performance.
2. Successful performance measures are based on a framework such as the BSC at the enterprise level.
   - Using small numbers of vital measures is a feature of the best performance measurement systems. By reducing the data available to the critical factors, people are able to focus on what is most important.
   - Financial measures are not the majority of measures. Adding a majority of non-financial measures is a consequence of relating to the strategic plan.

- BSC is used at the corporate or enterprise level (typically 10–20 measures). The BSC may also be used at other organizational levels, particularly when a business unit has a need to have its own strategic plan.
- A family of measures (five-to-ten measures) is typically used at business process levels throughout the hierarchy.
- The system usually includes predictive measures for the look ahead. The predictive measures are typically control factors because these activities can be built into business plans and will produce the desired performance.
- Direct measures of customer and employee satisfaction are usually included. The customer perspective is an important part of the BSC; however, a customer satisfaction index may not be a critical measure because of the difficulty associated with making the measurement and understanding the results. Employee satisfaction is also important for the internal business perspective; however, this measurement may not be viewed as one of the top five factors for success from the business perspective.

3. Wide-ranging education is required to develop and apply measures in the operating units.
    - Employees are involved in developing operational-level measures based on business processes. Involvement of people in the process creates ownership of the performance measures.
    - Employees can relate measures directly to their jobs. By being part of the way business is done, performance measurement is directly related to work, not an add-on.
    - Corporate function drives education and implementation throughout all operating units. A small number of people are needed to train employees on performance measurement. Without the proper training, many people will not get the full benefit from performance measurement.
    - Frequent review of measures with interpretation of results. Periodic review of results is typically a quarterly or yearly function for large enterprises. Business units within the enterprise will report results quarterly or monthly. Work teams may report performance measurement very often, ideally in real-time or near real-time.
    - Periodic refinement of measures and raising the bar (annual adjustments, target setting) is done to improve performance. Target setting is a difficult aspect of performance management and is one reason for concentrating effort on a small number of measures.

4. Systematic data collection/roll-up must be implemented through formal processes and systems.

- Regular/periodic data collection is made through various information sources and systems. The data required for strategic purposes are usually scattered among a number of systems and, in some cases, have not been automated.
- Employ a variety of technologies—many use homegrown solutions. The solutions may be built around a financial reporting application or from scratch.
- Formal, complex processes roll-up local data into corporate measurement systems. Variation in definition of measures within the organization often requires an interpretation process.
- Automation of data gathering needs further development. Many companies use computer applications that require significant manual attention or provide measures that are not on the BSC.
- Tracking performance over time shows trends and relationships (historical data). Trends are often as important or more important than the absolute measures because trends will show long-term improvement or problems.

5. Measurement results must be widely communicated and acted upon.
   - Formal, regular publishing of results via company newsletters, intranets, etc., gives everyone in the company the same picture of how the company is performing.
   - Regular, informal, in-person communication of measurement results (e.g., via emails, team meetings, etc.) is used to stimulate participation in improving performance or improving the performance measurement system.
   - Road show or high visibility communication by senior leaders shows employees their contribution is important and helps keep the organization aligned to the main performance factors.
   - Cause and effect relationships are analyzed for business process improvements. The control factors (or leading factors) may have a complex relationship with the performance measures. Refining this relationship enables better forecasting of performance.
   - Team involvement and responsibility in adjusting performance puts the emphasis on results. When done on an individual basis, performance measurement is often seen as a tool for personnel management and often results in outstanding individual performance even when the team or company performance is poor.

2.2 Performance Measurement Framework

One fundamental concept in understanding performance measurement that puts various measurements into perspective is the use of a performance measurement framework. A good framework identifies measures that tell to what extent the organizational vision and strategy is being achieved.

For this project, the BSC was selected and modified as the basis for a performance measurement framework for water and wastewater utilities. The BSC framework is based on the understanding that an organization's continuing success relies on more than financial factors alone.

The BSC used in this project contains performance measures from four perspectives (shown in Figure 2-1) developed to reflect a balanced view of a public service organization. This measurement framework, along with a business process framework as presented later, forms the basis for a performance measurement system.

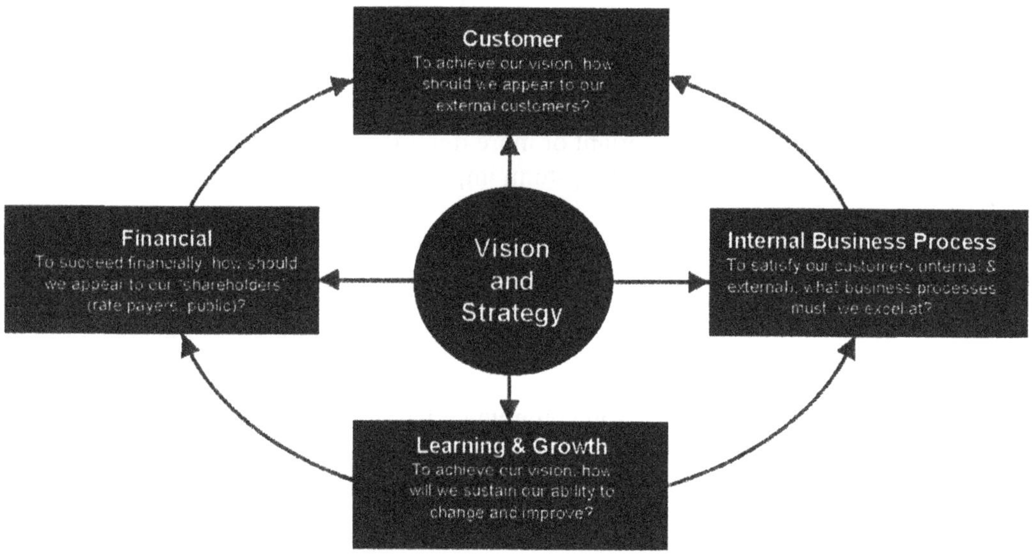

*Kaplan & Norton, 1996 (modified)

Figure 2-1. Most Widely Used Framework: Balanced Scorecard Was Used.

Historically, many businesses have been out-of-balance with respect to measures of performance. Driven by quarterly earnings and financial performance, financial measures have dominated the private sector business performance measurement systems. Although wastewater/water utilities may not be driven by the same financial performance reporting, they are driven by regulatory performance reporting. Again, this leads to out-of-balance emphasis on performance measures driven mostly by regulatory reporting requirements. An ideal system of measures will balance this emphasis and link measurement to an organization's vision, strategy, and goals.

Four perspectives give a balanced view of how well the utility is achieving its vision and strategy:

- ***Customer***—This perspective captures the ability of the utility to provide quality water and services including the effectiveness of their delivery and overall customer service and satisfaction. The focus is on identifying what the external customer values and expects.

- ♦ ***Financial***—This perspective emphasizes cost efficiency, (i.e., the ability to deliver maximum value to the ratepayers and public shareholders. Long-term fiscal solvency is based on optimized life cycle of all resources with sound, equitable rate support.
- ♦ ***Internal Business Processes***—This perspective focuses on the internal business results that lead to financial success and satisfied customers. Utilities identify the key business processes where they must excel in order to meet the utility's goals and objectives. Key processes are monitored and adjusted to meet performance objectives.
- ♦ ***Learning and Growth***—This perspective focuses on the employees, the quality of information systems, and the effects of organizational alignment to support accomplishing goals and objectives.

The generic measures that show up in most organization scorecards are summarized in Table 2-1. Each organization's specific measures are designed to support its particular strategic objectives.

Table 2-1. Generic Performance Measures. (Source: "The Balanced Scorecard," Kaplan & Norton)

Perspective	Generic Measure
Financial	Return on investment and economic value added
Customer	Satisfaction, retention, market and account share
Internal Business Process	Quality, response time, cost, and new product introductions
Learning & Growth	Employee satisfaction and information system availability

These perspectives are interrelated. The customer perspective is influenced to a great degree by how well the internal business processes are carried out and at what cost (financial perspective). Sustained performance of internal business processes and financial perspective are influenced by learning and growth of employees and the organization. These interrelationships are indicated in simplified form by the arrows in Figure 2-1 that indicate an *in order to* relation. These arrows indicate that the learning and growth perspective goals support each of the other perspectives. For instance, in the learning and growth perspective, the objective is to improve employee skills (possibly measured by targeted training hours per employee) in order to improve the quality of work (the internal business process perspective) and meet customer needs (the customer perspective). The ultimate achievement of the vision, strategy, and other perspectives is the customer perspective goals.

Each of the perspectives in the scorecard is defined in terms its goals, objectives, measures, and targets:

- ♦ *Goals:* Goals are end-state or results of accomplishing or carrying out the strategy. Goals are usually broader than objectives and may break down into several objectives. Each perspective typically has between one and three goals.
- ♦ *Objectives:* Objectives must be measurable (sometimes it is not necessary to break down goals into multiple objectives; in these cases, the goal and objective are the

same but must still be measurable). Each goal is usually further defined by two or three objectives.

- *Measures:* Measures are the quantification of meeting an objective in terms of effectiveness, quality, or efficiency (or activities if direct measure is not possible or practical). Objectives can have several measures, and picking the best measure is sometimes difficult and may change over time. Each objective should focus on one or two good measures using the best data available.

- *Targets:* A target is the desired value of the measure. Targets can be stated as absolute values or changes to current or baseline values. Initial targets can be set where they are obvious or clear or where baseline data are available. In other cases, targets may need to be set following initial data collection.

A sample portion of a BSC for the customer perspective of wastewater services is shown in Figure 2-2.

Customer			
Goals	**Objectives**	**Measures**	**Targets**
Customers See Us As Stewards Of The Environment	Compliance with NPDES Permit	% Compliance with NPDES Limits	100% Within NPDES Limits
	Eliminate Substantive SSOs by 2005	% of SSOs Reduced Via FOG Program	80% Reduction in Backups
		% of CMOM Program Components Implemented	Fully Implemented in 2005
Provide Customers Consistent And Timely Service	Readily Accessible Customer Information	Response to Customer Query	Within Four Hours of Initial Call
		Information Available Via Website	80% of Queries Answered Via Website

Figure 2-2. Sample Balanced Scorecard.

As shown in the sample scorecard, goals are expressions of the permanent aspirations of the organization, either currently being achieved or to be achieved in the long term. Objectives are further definition of a goal into results that are measurable. An objective must be measurable by one or more measures of effectiveness, quality, or efficiency. As shown in Figure 2-2, some objectives are activity-based (e.g., eliminate substantive sanitary sewer overflows by 2005) and will change with time.

Objectives for each goal are set that if met will lead to the goal being achieved. Measures are then established that show how well, or to what degree, the objectives are being met. Targets for each measure are set periodically (e.g., annually) to improve performance in line with moving toward stated objectives.

2.3 Business Process Framework

Another fundamental concept of performance measurement is to link the measures to business processes so it will be clear what actions must be taken to improve business performance. A business process framework is essential for developing a BSC because it links work processes to each of the scorecard perspectives.

The business process framework can be used to view performance measures from either of two directions:

- *From the top down:* Using an enterprise scorecard (e.g., utility-wide or corporate scorecard), the process framework can be used to verify which measures will apply to which areas of the enterprise. This verification ensures that the enterprise scorecard drives organizational performance by showing measures that apply to processes in the entire organization.
- *From the bottom up:* The process framework can be used to define process-based scorecards. These process-based scorecards must link and align with the enterprise scorecard through common goals and objectives. This alignment helps the entire organization be unified on common goals while the process-based measures are meaningful to employees who do the work. These process-based scorecards as well as initiative-based scorecards are further described later in this chapter.

The business process framework used in this project was developed based on the utility business architecture (UBA) (AwwaRF, 1997) and the American Productivity & Quality Center's (1999) process classification framework (Figure 2-3). The business process framework provides the structure for measures that are process-based for each of the scorecard perspectives. An explanation and further description of the utility business process framework is included in Appendix F.

The UBA is a generic process framework for a typical wastewater and water utility that shows its core services and business services. Core services are those that are delivered or add value directly to a utility's customers (usually its ratepayers and the public). Core services are typically provided by the organization or enterprise itself. (Changing the provider of the core service usually means changing the mission and/or vision of the organization.)

Business services are those services required by core services that support and enable the core service delivery. Business services may be provided internally (by the organization or enterprise itself), by a parent organization (e.g., usually a municipal or regional provider of services to several departments), or outsourced to other service providers, public or private.

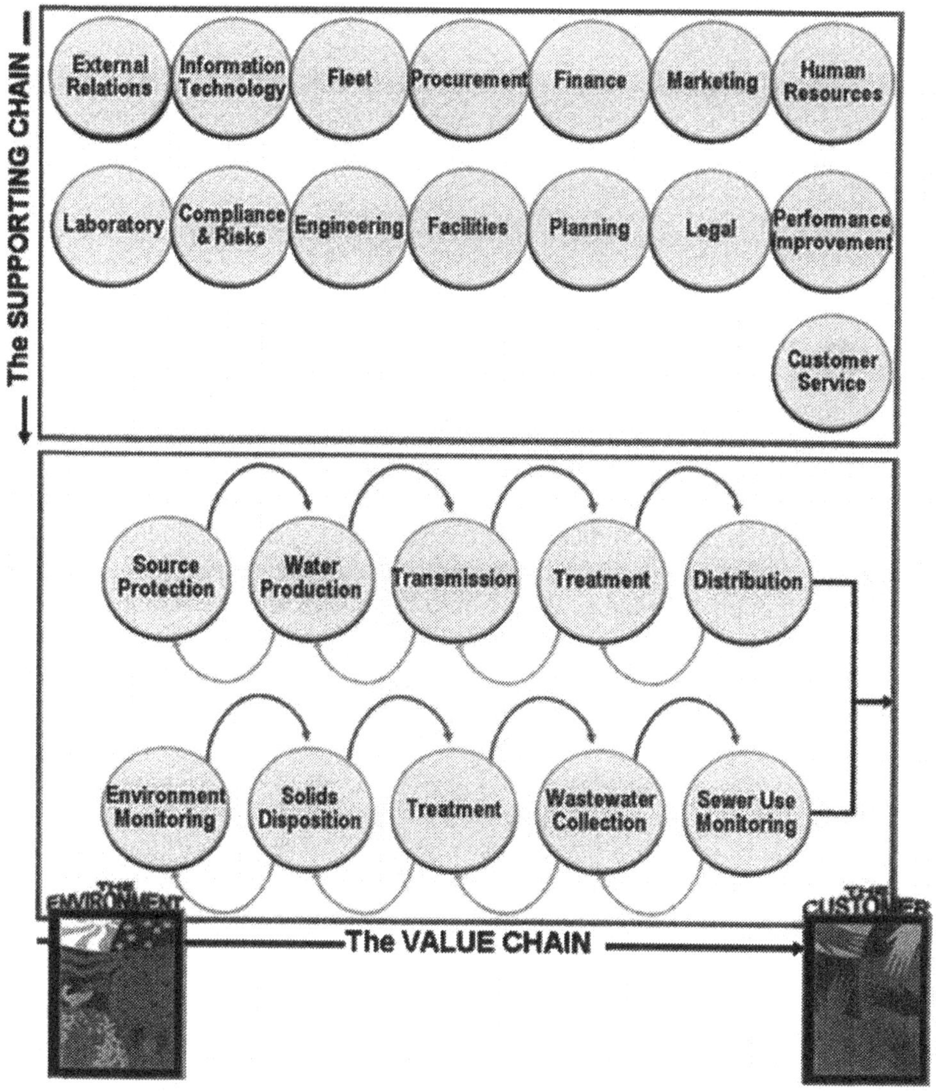

Figure 2-3. Business Process Framework Is Based on the Utility Business Architecture (UBA).

2.4 Types of Measures

Another fundamental concept of performance measurement is selecting measures of the right type. Some measures record what was done (cost, effort, number of work orders, or other activity), whereas other measures indicate results (amount of wastewater treated, percent of wastewater sent to treatment, quality of treatment). A good measure should be easy to understand and hard to falsify.

A scorecard should be balanced on measures of efficiency, quality, and effectiveness at each level of the performance measurement system, whether at the enterprise level, a process, or an initiative level. The types of measures include efficiency, quality, and effectiveness (Figure 2-4) as described below:

- ***Efficiency***—Ratio of inputs to outputs for a given process. For example, the amount of energy used divided by the amount wastewater treated is an indication of the energy efficiency of a treatment process or cost per volume produced (e.g., $/MG) inclusive of all costs (e.g., labor, chemicals, energy).
- ***Quality***—Meets an established standard. This is the extent or degree to which the process output meets the stated standards (standards can be defined by customer, regulator, governing body, or internally). All regulatory standards for water quality are quality types of measures. Quality measures could also be the percentage of activity performed meeting some utility-established standard, such as the percent of preventive-maintenance procedures completed on time.
- ***Effectiveness***—Outcomes or results. This is the extent or degree to which the process meets the needs of customers or accomplishes its purpose, outcomes, or results, such as employee turnover rate, percent of actions that result in customer service improvement, or number of water-borne disease outbreaks within a community.

Each type of measure is defined in order to create balance in the type of measure within each scorecard perspective or within a given business process or strategic initiative. The challenge is to be selective in defining measures to establish the smallest number of most relevant measures.

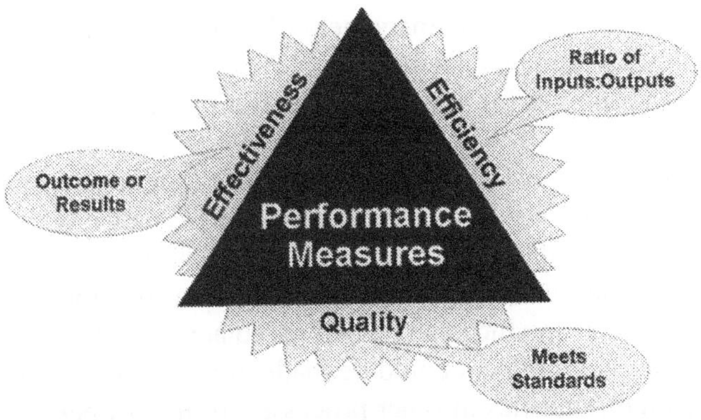

Figure 2-4. Good Performance Measurement Applies Three Types of Measures within Each Scorecard Perspective.

Another dimension of balance in the type of measures is leading versus lagging indicators.

- **Leading** measures are predictive of or possibly correlated to a process result or outcome, which can also usually be measured. As such, they are typically measures of process inputs or variables in advance of the process output or result. They can provide a basis for action to prevent an adverse outcome. For example, a vibration sensor on a large pump may be used to trigger a maintenance activity in advance of the pump either failing or needing to be shutdown in an emergency. The vibration measurement is a leading indicator of the asset condition.

- **Lagging** measures are after the fact and typically are measures of process outputs or results. While these measures are often too late to take corrective action, the trend, charge, or rate of change in these measures can often be used as a leading indicator.

Sometimes measures that are lagging in one process can be leading in another process. For instance, a measure of information system network availability or throughput may be a lagging indicator for information technology (IT), but a leading internal business process indicator for IT customer service or user satisfaction.

2.5 Process-Based Scorecard Development

A benefit of process-based scorecards is that they show value-added processes from a customer's perspective, as well as relating employees' work to the process goals, objectives, and measures. A utility providing wastewater collection and treatment for customers has a set of core processes used directly to collect and treat wastewater. These processes add value to wastewater generated by customers and create the motivation for customers to buy services from the utility. Customers expect that wastewater will be collected at all times, transferred entirely to treatment, and treated to safeguard the environment. They are interested in the performance of these processes or services. They are also interested in the cost. To get the full cost of service, it is necessary to add the cost of applicable indirect business processes (such as human resources, financial, legal, and planning). Customers are also interested in how these business processes are performed; for example, they expect fair management of workers, sound financial position, good legal protection, and timely investments. By relating performance to customers, the utility will gain their understanding and support for new initiatives and improving performance.

Another benefit of the process-based scorecard is that it defines typical water/wastewater utility functions, which in turn allows utilities to compare performance with other utilities on a common basis. For example, the laboratory process is defined as a separate collection of processes such as sampling, analytic and qualitative testing, quality control, and test result analysis. Laboratory process performance can be compared across utilities for quality, efficiency, and effectiveness. The latter measurement could be made either in cost per million liters treated, which can be rolled up with efficiencies of other processes, or in cost per test for each type of test or groups of tests. Measures of effectiveness could include timeliness of getting results to all users. Measures of laboratory process quality could include the accuracy of test results based on conformance to calibrated laboratory standards or annual laboratory certification results.

The framework not only provides the basis for process-based performance measures, it also provides the basis for practice comparison. In the case of the laboratory process, the pros and cons of a single, central laboratory versus plant laboratories could be evaluated. Other practice comparisons could be made for the chain of custody and tracking of samples, scheduling of batch tests, and information management.

The focus on process addresses all efforts required to provide a service, cutting across organization boundaries as needed. Process-based scorecards are not tied to the organizational structure; so cross-functional teams are used in developing and implementing them, breaking down organizational silos. Utilities using this approach will find that departments and employees will better appreciate their role in providing services, giving them additional motivation in working toward utility goals. Utilities may also find opportunities to improve the processes, organization, or other aspects of getting the work done.

In a process-based scorecard for water services, examples of measurements for source protection and water distribution in the water value chain are shown in Figure 2-5.

	Source Protection	Distribution
Effectiveness	No. of Source Contamination Incidents	No. Tap Water Related Disease Outbreaks
Quality	Influent Quality within Treatment Constraints	% SDWA Regulations Met
Efficiency	Total Cost/ Acreage Protected	Total Cost/Volume Delivered ($/MG Distributed)

Figure 2-5. Example Performance Measures for Water Value Chain Services.

Water source protection should provide raw water to production in sufficient quantity at a quality that can be treated to provide acceptable drinking water. One measure of how effective a water source is being protected is the number of times that the source is contaminated from various incidents. For example, a raw water reservoir fed from a river could be contaminated from a chemical spill into the reservoir or within the watershed of the river upstream of the abstraction point. In the former case, the utility could well have total responsibility for preventing such contamination, whereas in the latter case, the utility may have very little responsibility. Nevertheless, the number of source contamination incidents is a good measure of how well it is being protected.

One measure of water distribution effectiveness is the number of outbreaks of disease due to impure tap water. The outbreaks could also be measured as the number of people affected. Some outbreaks could be due to improper treatment rather than infection during distribution.

In a process-based scorecard for wastewater services, examples of measurements for wastewater treatment and sewer usage in the wastewater services value chain are shown in Figure 2-6.

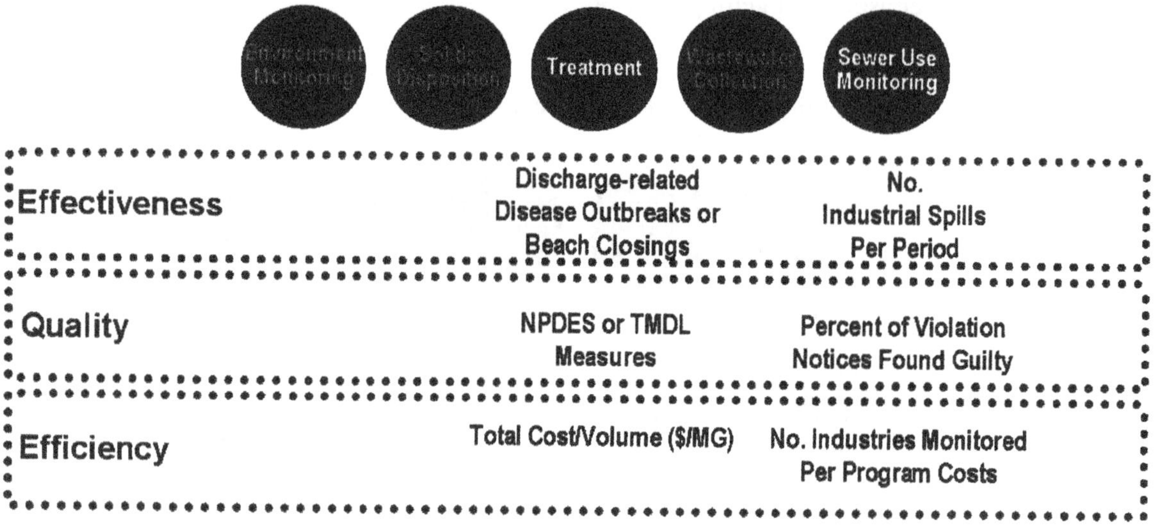

Figure 2-6. Example Performance Measures for Wastewater Value Chain Services.

The number of industrial spills into the sewer system may be partially outside of the control of the utility; however, an effective program of sewer usage monitoring will reduce the number of incidents due to intentional spills.

Examples of measures of wastewater treatment are the number of discharge-related outbreaks (or people affected) of disease and the number (or number of days) of beach closings. Eliminating such outbreaks is one of the primary reasons for treating wastewater. Quality of treatment can be measured by the quality of effluent either in absolute terms or as a percentage of discharge limits. By looking at effectiveness, quality, and efficiency, the view of performance is balanced between outcomes and effort.

2.6 Initiative-Based Scorecard Development

An alternative to process-based scorecards that is also bottom-up is an approach using initiative-based scorecards. As part of strategic direction, many organizations have a number of key or strategic initiatives that can be the basis for scorecard development. Like process-based scorecards, they must also link and align with an enterprise scorecard. Initiative-based scorecards may also help employees improve performance through the initiative and relate to meaningful measures.

As an example of an initiative-based scorecard, the importance of balanced performance measurement can be seen particularly well in the maintenance service areas. The effects of maintenance often have a significant time delay from when maintenance is done. For some equipment and facilities, it can take several years of under-investment in maintenance to show itself in poor performance of the equipment and facilities. Therefore, maintenance performance is a long-term balance of effectiveness, efficiency, and quality. Examples of measures for managed maintenance are shown in Figure 2-7.

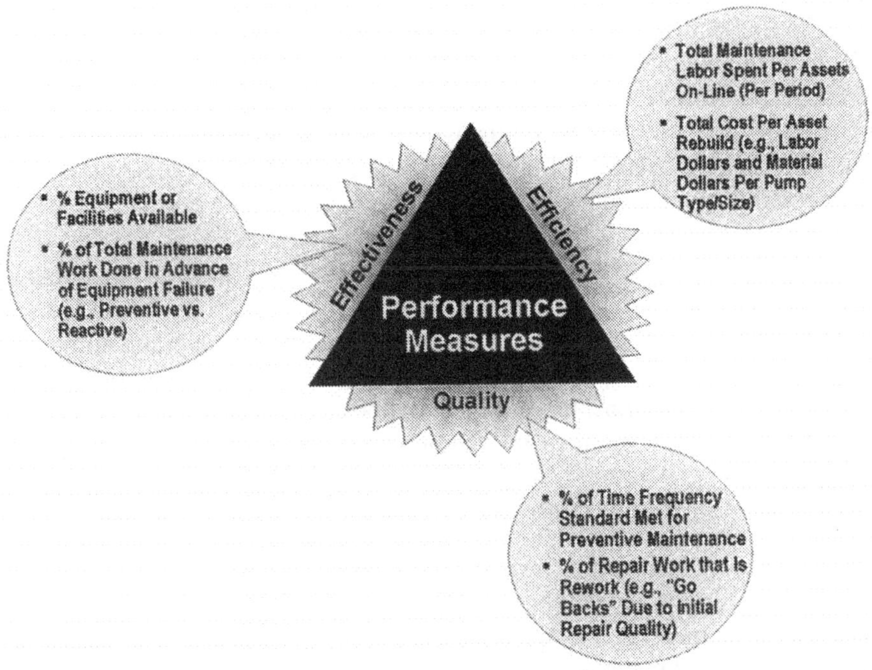

Figure 2-7. Examples of Measures for Initiative-Based Scorecard in a Managed Maintenance Program.

Effectiveness of maintenance must be viewed by how maintenance affects the services that rely on working equipment or facilities. Effectiveness can be measured in the number of breakdowns or the percentage of time that equipment and facilities are available for use. This measure is always valid for reporting effectiveness, but it may not be sufficient for all assets.

For example, a length of sewer may be continuously available for service for many years without any maintenance. In this case, the availability of the asset is not a good measure of how effective is the maintenance. For this type of asset, a measure of maintenance effectiveness could be the percentage of maintenance work done in advance of asset failure (i.e., the percentage of preventive maintenance versus the percentage of reactive maintenance). Such a measure may show how effective a program of sewer inspection and replacement is at preventing sewer collapse. However, this measure must be used with a great deal of caution because it represents an activity rather than the actual result. It is very useful for controlling maintenance budgets with a long-term view of achieving effective maintenance service, but the percentage of preventive maintenance is not an outcome of maintenance work.

Quality of maintenance is measured by comparing the results of maintenance with a reference or standard. A quality measure could be the percentage of time that the frequency standard for preventive maintenance is met. For example, if preventive maintenance of a fire alarm system is to be done quarterly, but is actually only performed three times during 1 year, the quality of preventive maintenance would be 75%. The percentage of maintenance work orders that have to be redone because the original work was not done properly is another measure of work quality. In this case, the standard or target for rework is kept separately.

Efficiency could be measured as the total cost of maintenance labor in a year divided by the total value of assets maintained. Maintenance labor includes work done by the maintenance staff, operations staff, and contractors. Asset value should be the replacement value. This measure of efficiency is very useful in planning asset replacement based on economic life of the asset; it is also used to determine areas of maintenance practice that should be re-examined. This overall efficiency could be measured monthly, quarterly, or other time period.

Another example is the total cost of maintenance (labor and material) in a year divided by the total value of assets maintained. Efficiency can also be measured for specific tasks; for example, the total cost of a major maintenance task such as rebuilding a pump compared to the value of the pump. The ratio of pump rebuild cost as a percentage of pump value can be tracked for different types and sizes of pumps. It would then be possible to determine the economic frequency of performing a rebuild for the range of pumps being maintained.

2.7 Shifting to a Measurement Culture

Shifting to a measurement culture, where the benefit of measuring performance becomes sustainable over the long term, is a challenge in organizations unaccustomed to measurement. There is little point to implementing a measurement system if the benefits will not be realized. Performance measurement and management should be:

- A way to align the whole organization on what is important;
- A system for increasing organizational accountability;
- A commitment to change; and
- A way to balance what is important to all stakeholders.

Performance measurement and management is a process or journey, not a one-time task. The cycle of planning, doing, checking, and acting should be ongoing to keep the targets and measures meaningful to the business and a relevant part of daily operations.

Successful case histories implement a strategic organization-wide performance management process rather than a tool to measure and control individual productivity. A focus on the organization's performance will allow individuals to align their goals to achieve the greater goals of the organization. As part of the process, it is extremely important to communicate company performance to everyone.

For an individual, a line of sight to the measures, objectives, and goals of the organization is key to alignment of the entire organization on its strategic direction. Individuals or teams (such as process-based teams or initiative-based teams) should see the relevance of their efforts to the measures as directly as possible. Measures that are meaningful at an individual or team level should be seen as part of and contributing to measures at the enterprise level.

By development of a performance measurement system specific to a utility, the utility has control over its performance. Although various published measures might be good candidates for a BSC, each utility must determine the important measures to achieve its vision. Published systems of performance measurement are not a fixed answer for every utility's measures. However, where published measures such as QualServe or industry regulation measures such as OSHA or U.S. EPA standards fit the appropriate goal and objective, they can provide an additional comparative measure to other organizations that use the same measures.

Shifting to a measurement-based culture takes time, enthusiasm, and commitment and requires:

- Visible leadership at senior levels. Visible leadership shows commitment and is needed not only at the enterprise level, but also as performance measurement and management is applied to business processes and initiatives throughout the organization. This includes specific assignment of responsibility to coordinate and support the data gathering and analysis and move the process forward.
- Employee involvement in developing and implementing measures. This builds staff commitment and educates the organization about its vision, goals, and services as well as performance management.
- Frequent, periodic review of measures and results. This reinforces commitment and builds enthusiasm as the organization achieves balanced improvement in its performance.
- Communication at multiple levels with multiple methods. This is needed to reach all staff.
- Adjustment to measures, targets, and business processes to improve performance. Targets set in an interactive process between management and staff is a critical part of building enthusiasm and commitment. It also takes several cycles of measurement and target setting for everyone to gain enough experience in how the process works, see how they fit into the organization's improvement, and see sustained, meaningful improvement.
- Linking measures and results to team-based incentives. This ensures that staff members receive direct feedback on their performance.

A seven-step process for developing and implementing performance measures in a measurement-based culture is described in Chapter 3.0.

CHAPTER 3.0

DEVELOPING AND IMPLEMENTING A SYSTEM

3.1 The Purpose of a System

The purpose of a performance measurement system is to:

- Align efforts of the organization and its employees to achieve a commonly held mission and vision;
- Report clear, concise measures of performance in context with organizational goals, objectives, and targets; and
- Improve quality, efficiency, or effectiveness of services delivered by the organization by acting on the feedback that measures provide.

In its fullest sense, the system of a performance measurement system is the process by which utilities continuously improve their business performance. The performance measurement system begins with or flows from a utility's strategic plan. The strategic plan is where the responses to competitiveness, quality, and effectiveness issues are put in context. System implementation starts with defining the strategic plan and ends with using performance measurement in daily operations. The system is as much communicating the utility's performance as it is the behind-the-scenes data-gathering procedure and tool. A successful performance measurement system requires the full system, inclusive of processes and practices, organizational understanding and culture, and supporting data and technology.

The development of appropriate, useful performance measures that truly improve business results is an important component of a performance measurement system. Defining and selecting a balanced set of measures that link to goals and objectives based on business strategies is the subject of a significant part of this report.

3.2 Overview of Seven-Step Process

Common Process for Developing and Implementing a System

This chapter provides a process for establishing performance measures within the perspectives of the BSC approach and for ensuring those measures fit within an overall management approach.

A seven-step process for developing and implementing a performance measurement system was created from best practice information and other secondary research completed during Phase 1 of the project. Completion of each of these seven steps is essential to successfully developing a performance measurement system that will result in improved performance (Figure 3-1). Each of the steps will be described further in this chapter.

Figure 3-1. Seven-Step Process Provides Common Approach for Developing and Implementing a Performance Measurement System.

Step 1. Review the Utility Strategic Plan

Utility examples demonstrate that linking a performance measurement system with a utility's strategic direction provides organizational alignment (Figure 3-2).

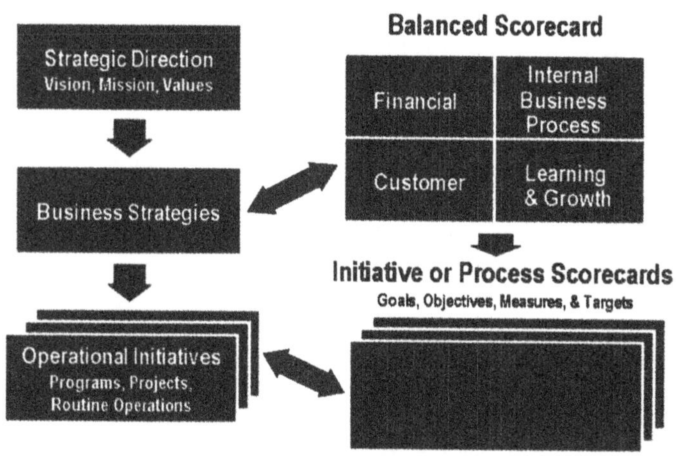

Figure 3-2. Performance Measures Must Align With Utility's Strategic Direction.

Alignment of business strategies to support the strategic direction should extend to programs and routine business processes that carry out the work of the organization. By aligning the efforts of the organization, maximum efficiency should be achieved. Organizational alignment means that all staff members know what is important to the organization, how their position fits into the overall framework for providing services, and what they can do to improve performance. The experience that defined best practices in Chapter 2.0 suggests that organizational alignment is important in managing performance.

This step begins with a review of the utility's vision, mission (including customers and provided services), and business strategies to accomplish the vision and mission.

A utility's strategic direction, business strategies, and related initiatives must provide the goals and objectives for a performance measurement system. This framework means that enterprise and business scorecards are aligned top down with strategy and bottom up with program, initiatives, and on-going operations and business processes.

Each of the utility's operational initiatives or major business purposes that carry out the business strategies can be measured by team-based scorecards. This utility-wide system of measures with related scorecards can be used to apply performance measurement on an enterprise basis, with meaningful team-based measures that link to the utility's goals and objectives.

Performance measures for a utility's initiatives show how these initiatives will directly carry out and support a utility's strategic direction. The initiative's objectives, measures, and targets also provide feedback on performance directly to work teams or individuals. With systematic performance feedback, these teams and individuals can make operational improvements. With a cycle of continuous improvement, utilities can reach new levels of performance for customer and public benefit.

Step 1A. Review Utility Vision, Mission, and Strategy

This methodology requires a utility vision, mission, and strategy to be in place first. If these do not exist, then the organization should at least create a working version of a vision and mission to proceed with the process. This ensures that performance measures developed in each perspective support accomplishment of the utility's strategic direction. Development of vision and mission statements facilitates identification of key business strategies. Business strategies translate a utility's strategic direction into operational initiatives.

An example of a utility's business strategies and their emphasis in each scorecard perspective is shown in Figure 3-3. This matrix also can show the balance of business strategies.

Scorecard Perspective Emphasis

Business Strategy	Customer	Internal Business Process	Learning & Growth	Financial
1. Protect, sustain, and renew cost-effective water resources	●	○		○
2. Provide exceptional customer service	●	○	○	
3. Protect the public health and are stewards of the environment	●	○	○	
4. Practice sound financial management		○		●
5. Foster trust and empowerment of stakeholders through open and timely communication and information sharing	○	●	○	
6. Attract, retain, and reward a highly capable workforce		●	●	
7. Optimize our core and support/ business services	○	●	●	○

Figure 3-3. Balanced Scorecard Adds Perspective to a Utility's Business Strategies.

Step 1B. Identify Operational Initiatives

Operational initiatives, such as programs, projects, and on-going operations as shown in Figure 3-2, carry out high-level business strategies. By preparing an inventory of operational initiatives, a utility can correlate initiatives to the business strategies they help carry out. Typically, a strategic plan would already define these initiatives and relationships. If an initiative cannot be matched to a business strategy, either it is not a significant initiative and therefore not part of strategic direction, or else not all business strategies have been defined. Some business strategies developed to support strategic direction may not have any existing operational initiatives to support them. Initiatives should be developed or created to support all business strategies.

Step 1C. Develop Goals and Objectives

Based on its business strategies an organization should identify goals and objectives for each scorecard perspective. The BSC translates a utility's business strategies into a set of performance goals and objectives for each of the four perspectives (Figure 3-4). After determining the need to improve in Step 2, measures and targets can be developed for each goal and objective in Steps 3 and 4. The output of these steps is the development of the enterprise BSC.

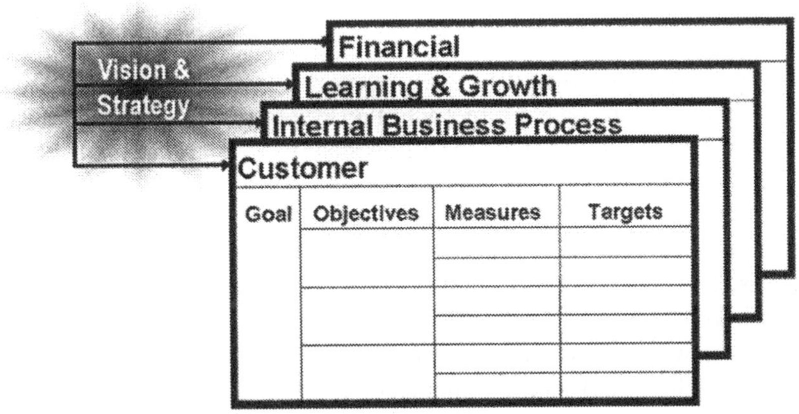

Figure 3-4. Balanced Scorecard Translates Vision and Strategy into Goals and Objectives.

Step 2. Prioritize the Need for Improvement

Areas for improvement and potential benefits (e.g., cost savings, improved quality, better service) must be defined so that the performance measurement system serves a purpose. Improvement areas and benefits are typically defined through a competitiveness assessment, benchmarking process, or review program such as QualServe.

Possible improvement needs include:

- Improving service effectiveness, quality, and/or efficiency;
- Aligning staff with organizational vision, mission, and business strategy;
- Increasing accountability to customers and the public;
- Driving change in business practices;
- Creating incentive structures for work teams and individuals; and
- Providing feedback on individual team and organization-wide performance.

The need for improvement provides the motivation for managing performance. Quantifying the need for improvement is essential for selecting good measures (Step 3) and setting targets (Step 4). For example, a performance assessment may reveal a need to improve services and to reduce costs. The need should be quantified where possible and the measures used to assess performance should be considered for inclusion in the performance management system. Improvement may be in reference to previous performance by the organization or in reference to an accepted standard or other external reference.

Measurement against performance of other similar organizations may be interpreted subjectively. That is, the organization's performance may be rated as equal to the average of similar organizations, but if the goal of the organization is to be in the top 10%, there is a need to improve.

The process for determining the need for improvement may also change the objectives and goals defined in the previous step. It may even challenge the expressed vision and mission of the organization. After determining need, re-examine the output of Step 1 and if necessary, refine the strategic direction.

Step 3. Define the Performance Measures

Performance measures are linked directly to the strategic direction of the utility as defined by goals and objectives for each scorecard perspective. Measures focus on the outcomes necessary to achieve those goals and objectives.

A list should be developed of business processes that make up or contribute to the operational initiatives. This task extends organizational alignment further, putting more of the organization's work into the overall plan of achieving strategic direction. Aligning business processes with initiatives will also reveal performance targets and measures that should be considered in developing the performance management system. It is not unusual for a utility to discover some business processes that do not contribute toward any part of the strategic plan. These processes should be scrutinized closely to determine whether they should be discontinued.

The appropriate measurement type (e.g., quality, effectiveness, or efficiency) should be determined for each directly applicable goal/objective. Clear, written definitions for each measure are needed so that everyone agrees on what is being counted. Top priorities should be selected for each scorecard perspective. Specific measures should be identified for each priority goal/objective. Results must be reviewed to ensure there is a balance between number of measures per perspective and measurement type.

The method and frequency of taking measurements must be determined. A responsible person or group should be assigned to gather and track measurements. The source of the measurement data (e.g., maintenance data, SCADA system, operator's logs, human resource records, financial reports) should be identified.

Step 4. Establish the Targets for Improvement

Based on the performance measure defined, a target should be set for each measure related to the need for improvement from Step 2. Target setting is sometimes obvious, especially when goals and objectives have clearly desired outcomes. Examples of obvious ones include an objective of eliminating sanitary sewer overflows, where the measure is the number of overflows per time period (e.g., month), and the target may be set at zero for major overflows (e.g., >1000 gallons). Because a target of zero overflows may not be practical, it may be better to set an improvement target, such as a 50% reduction in number of overflows. Target setting is not always easy, especially when there is no baseline data to know the historical values of a newly selected measure. Different methods to set targets according to the type of target are presented in Table 3-1. Even with benchmarking data available, a utility will still need to select its own targets (e.g., top quartile, best-in-class, above average).

Table 3-1. Methods for Target Setting.

TARGET TYPE	USE WHEN	EXPRESSED AS...
Absolute	❖ Established Standards Exist - Industry - Utility Specific ❖ Benchmark Data is Available - Internal - External ❖ Other External Standard Exists - NPDES Limit - Customer Expectation - Service Level Commitment	❖ Number ❖ Percent
Range	❖ New Measure; No Baseline Exists ❖ Want to Drive Incremental Improvement	❖ Number ❖ Percent
Degree of Change	❖ Drive Incremental Improvement Over Historical Performance ❖ Measure Defined as Increase or Decrease ❖ New Measure; No Baseline Established ❖ Customer Expectation of Performance Improvement	❖ % Increase ❖ % Decrease

Target setting needs to fit into the goals of the organization and also be realistic. For example, the organization may have a goal to be recognized as an outstanding protector of the environment. As part of achieving this goal, targets are set for effluent quality that are better than required by permit. The long-term average suspended solids in the effluent has run between 14 and 19 mg/L. This performance may be close to the natural limit of the treatment processes being used with little room for improvement by changing operation practices. Setting a target of 10mg/L may align with the goal of the organization but may not be achievable by improved operations. Although review of practices may reveal a way to improve performance, a review of the treatment processes will likely be required to understand how sustained improvement can be achieved. Setting a more aggressive target is only effective if suitable treatment processes are in place and operated properly—a combination that can actually deliver the desired performance. While targets for physical, chemical, or biological processes may be constrained by the process itself, the same kind of constraints may exist in business processes. Therefore, achieving new levels of performance in some cases will require process redesign or re-engineering.

The performance assessment done in Step 2 will usually report the gap between current performance and some standard of performance. In setting targets, the plan for closing this gap must be mapped over time. Targets may be set in conjunction with initiatives or milestones in a long-term plan of improvement. For example, a 10% improvement in the time to respond to customer complaints may be linked to getting a new call center up and running, another 10% after a comprehensive training program is completed, and a final ten percent when additional vehicles or staff are ready. Such a progression in performance improvement could be planned over several years. Targets should be specific and achievable and have a set time period. For some measures, knowledge of statistical process control principles can help in analyzing trends and setting targets.

Step 5. Implement the Performance Measurement System

The necessary data collection, recording, manipulation of data, and presentation of measurement results by periodic reporting or an online dashboard must be established. It may be necessary to initiate new standard operating procedures or training material to put the system in place. The next steps are to identify implementation team members and provide training to the team. Periodic reporting and monitoring schedules should be established and carried out by the organization's managers and leaders. Finally, processes must be developed for taking corrective action when targets are not being met or progress is not being made.

Automated performance measurement is preferred because the process is accomplished without requiring people to disengage from their normal duties to perform what might be regarded as an unimportant task. Tracking of performance over time allows a feedback system to visibly display measures as shown in Figure 3-5.

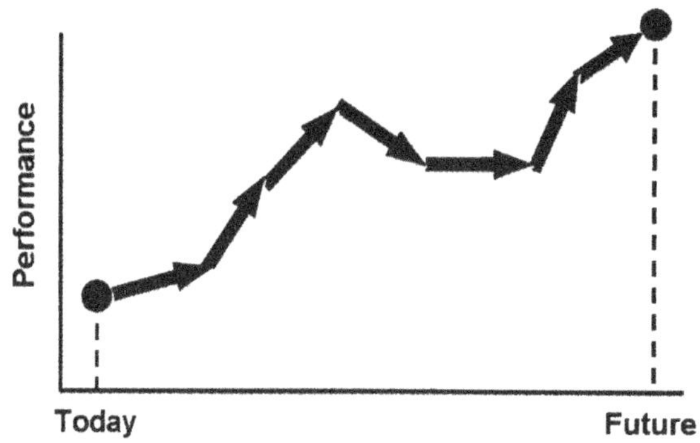

Figure 3-5. Performance Improvement Requires a System to Measure the Results of Changes—and Continuously Improve.

The development of appropriate, useful performance measures that truly improve business results is an important component of the measurement system. Therefore, discussion on defining and selecting a balanced set of measures that link to strategies, goals, and objectives is a significant portion of this report.

The measurement information provided to each user (or team of users) should enable them to adjust the business processes for which they are accountable (as shown in Figure 3-6). The measures should be part of a system of measures that are related so that all users can see or understand the impact of their business process performance on the rest of the utility. This *so-that* chain (or *in-order-to* linkage) creates the relationship of measurements to tasks and activities that are meaningful throughout a utility. By showing the relationship of measures to tasks and activities that support a utility's higher-level goals and objectives, measures can be created that are meaningful both at the top and on the ground.

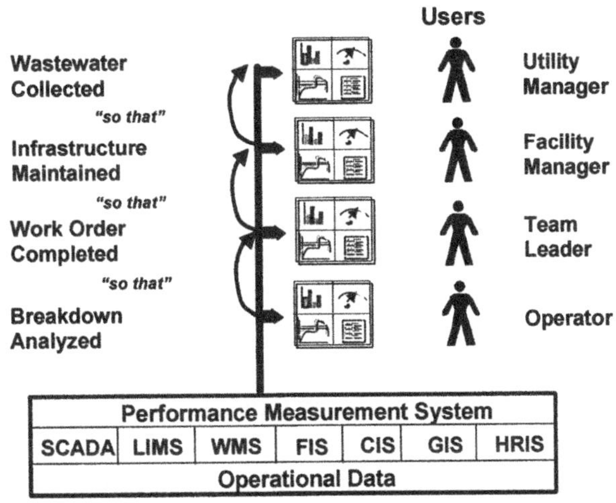

Figure 3-6. Example Measures Show Relevant Information to Each User as Feedback for Actions.

As also shown in Figure 3-6, a performance measurement system must typically consolidate and integrate operational data from a number of sources that can include automated information collection such as supervisory control and data acquisition (SCADA), laboratory information management system (LIMS), work management system or maintenance management system (WMS), financial information system (FIS), customer information system (CIS), geographic information system (GIS), and human resource information system (HRIS), as well as others. Although the performance data may not always reside in existing information systems, some degree of automated data collection will improve the success of implementing performance measurement.

Performance information originates in a number of different data sources throughout the utility that may be contained in various front-end information systems. For example, collection and treatment process information already is collected by a SCADA system. These systems typically have some amount of summary rollup data, as well as short to medium term trending capabilities. Many utilities have WMS or computerized maintenance management system (CMMS). These systems have data related to labor and inventory, including planned, scheduled, and completed work in terms of work content, hours, and costs. They also typically have inventory data. Other systems that many utilities have include LIMS to provide laboratory sample data, FIS to provide accounting related data, CIS to provide data related to utility customers, GIS to provide spatially-based information such as the location and relationship of facilities and utility infrastructure on maps, and HRIS to provide employee data such as training and development certifications.

All of these computer applications already have data collection, input, export, and reporting processes in place. Thus they are a rich source of the data needed by a performance measurement system. It makes no sense to duplicate the existing systems; rather the approach is to increase the use of the existing processes and data and only to supplement where required.

The performance measurement system is closely related to other business applications. It does not replace or supplant these systems, but leverages the information in them and is in fact deeply dependent on their existence.

Step 6. Measure and Initially Adjust

Following initial use of the performance measurement system, the measures targets and/or the systems that collect and present these measures will need adjustment from initial user feedback. Ask the questions:

- Are the measures the right measures?
- Are they well defined for accuracy and repeatability?
- Are the targets realistic?
- Are processes in-place to consistently gather the information?

It may be necessary to go back to Step 3 and 4 and refine the initial set of measures or targets.

At this point in the development process, it may be helpful to relate the seven steps back to the continuous improvement cycle. Referring to the Deming cycle presented in Chapter 1.0, Figure 3-7 shows how the PDCA cycle fits the steps.

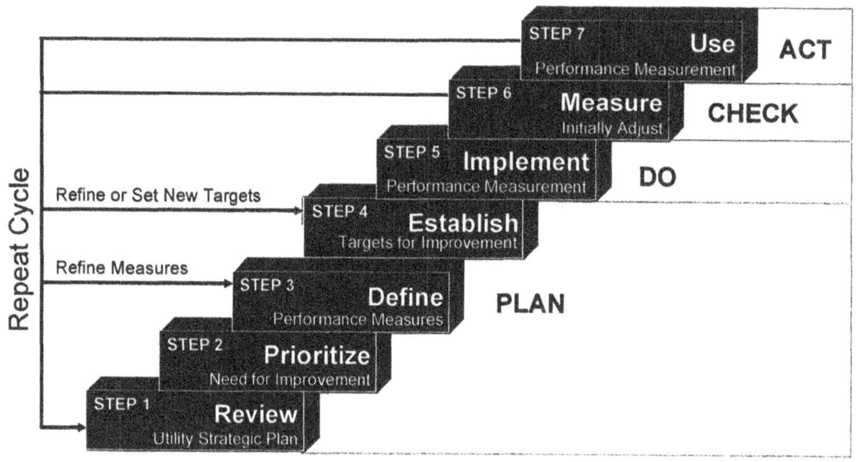

Figure 3-7. Steps for Developing and Implementing a System Are Related to the Continuous Improvement Cycle.

The first measurement period after a major definition will be slower and take more effort than subsequent periods. Training and analysis in groups should be promoted in order to make sure that people understand the system and its benefits, and also to provide feedback on any adjustments or changes.

Step 7. Use the Performance Measurement System

How will the utility know when it has an effective performance measurement system in place? Unfortunately, there is no magic button that pops up when the performance measurement system is finished. Continuous improvement and refinement is an ongoing element of a performance measurement system that requires a long-term commitment. It is important to review, refine, and/or revise measures and targets at least annually to improve the measures in terms of relevance, quality, and consistency or to raise the performance bar. If measures are changed, the loss of tracking or trending must be weighed against the better measure and correlation or translation may be needed for historical comparisons. Even so, a good relevant and actionable measure is better than just: "We've always measured it that way."

Full alignment throughout the organization means that scorecard goals and objectives apply to each area of the organization, including to the front-line operational teams. Senior leadership must set strategic direction of the organization and the system for managing its performance. It is also important that senior leadership is visible to the whole organization, particularly as the performance measurement system is rolled out to specific business processes and initiatives. During rollout, it is important to:

- Involve employees from all areas included in developing and implementing measures; and
- Communicate the strategic plan and related strategies or initiatives at multiple levels with multiple methods throughout the organization.

Regular use of the system means users take action based on the measures and periodically reset the targets to new levels of performance. Use of the performance measures in an ongoing improvement process or periodic review is critical. Historical trending and analysis should be performed where action is taken based on the actual versus target measures. If a target is not being met, the questions should be asked:

- Why not?
- Are there processes, technology, training changes that need to occur to meet the target?
- Or, is the target realistic?
- If not, what is an appropriate target?

If a target is consistently being met, does a more stringent target need to be met to encourage process improvements and growth? The importance of this task is that processes are periodically reviewed, assumptions questioned, and changes made where appropriate.

Simply providing the data to employees is not enough, though it is an important first step. The next step is to provide training to appropriate employees to help them make process improvements properly. Most employees may not be familiar with a performance measurement system concept or the types of data collected before. They need assistance in analyzing and interpreting the data so they can understand what it means. Once they understand the data and how it applies to processes for which they have responsibility, they will need help in developing and selecting alternative actions to improve the process. An ongoing education and communication program is required throughout the organization whenever a performance measurement system is implemented.

CHAPTER 4.0

ADJUSTING THE PROCESS

4.1 Measuring at the Top and on the Ground

From experience with the utility pilot projects as well as other organizations' use of the BSC approach for performance measurement, strengths and weaknesses of different ways of implementing scorecards have emerged. The body of knowledge available to utilities has grown considerably in the last several years, especially as more public sector organizations have embraced performance measurement. A number of references and websites are listed in the appendices that show that the BSC has become the predominant framework for performance measures.

In implementing scorecards, one of the major challenges is the relationship between measures at an enterprise level meaningful to executive management and at a process or activity level meaningful to front-line employees. A common approach to creating more meaningful and useful measures is the use of multiple scorecards throughout an organization, both across organizational units (e.g., divisions, departments, units, or groups) as well as up and down an organization or process hierarchy. While this approach is helpful for measures to be more relevant, it complicates the relationship, roll-up, integration, and consolidation of measures throughout an organization.

To contrast the pros and cons of enterprise, utility-wide level scorecards, and on-the-ground level scorecards (process- or initiative-based), the following summary is relevant.

Enterprise Level Scorecard

Pros	Cons
• **Align with strategic direction (vision, mission, goals)** • **Few key measures for the whole organization** • **Communication tool for customers, employees, and other stakeholders** • **Focus on "What's Important"**	• **Not directly meaningful to most of the staff – "doesn't relate to my job"** • **Just a way to provide incentives for top management** • **"I have no control over those measures"** • **Focus on corporate/financial measures** • **"What's in it for me?"**

On-the-Ground Level Scorecard

Pros	Cons
• Meaningful to team or individual performance • Measures that are trackable, day-to-day • Provides basis for team rewards/ uncertainties • Feedback for adjusting business process or practices (change the way/work)	• Different measures throughout the organization – how to consolidate? • Don't relate to strategic direction • Single perspective – not balanced • Inconsistently implemented by areas – some do, some don't

To address the shortcomings of using either of these approaches alone, the utility pilots demonstrated ways to use a combined approach. A combined approach will achieve these results:

- Align the enterprise scorecard with strategic direction
- Align team-based scorecards (e.g., process-based or initiative-based) with the enterprise scorecard
- Relate/roll-up ground-level measures to enterprise measures

To create a system of utility-wide measures that are related and link to a utility's goals and objectives requires starting with the vision and mission. As shown in Figure 4-1, the business strategies that support the utility's strategic direction can be tracked by an enterprise scorecard of performance measures.

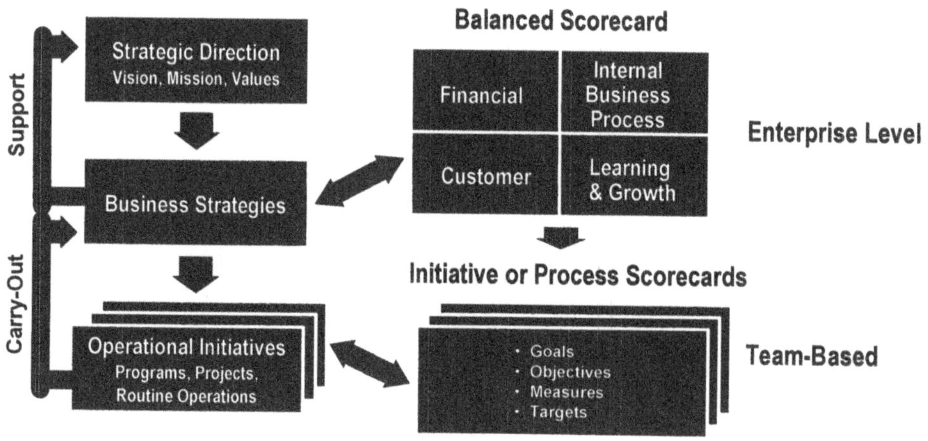

Figure 4-1. Scorecards Can Measure and Relate Performance at the Enterprise as Well as Team-Based Levels.

In addition, each of the utility's operational initiatives that carry out the business strategies can be measured by team-based (process or initiative) scorecards. Processed-based scorecards are linked to a business process framework that defines typical water/wastewater utility functions such as the UBA. Initiative-based scorecards are linked to an organization's key or strategic initiatives or programs, typically through a utility's strategic planning process. This utility-wide system of measures with related scorecards can be used to apply performance measurement on an enterprise basis, with meaningful team-based measures that link to the utility's goals and objectives. The challenge of relating and integrating this combined approach of enterprise and team-based levels can be addressed in many ways, several of which were applied to the utility pilot projects. As an example, Figure 4-2 shows the linking and overlap of scorecards in terms of hierarchies at the enterprise-level and team-based level for initiative- or process-based scorecards. Using this approach, the goals and objectives of both the enterprise scorecard and all team-based scorecards in an organization can be the same. The measures may be different, depending on the level and process or initiative area where those goals and objectives are applied. If the measures are different at the team-based level from the enterprise level, they should support, relate, or roll-up to the enterprise measures so that the enterprise goals and objectives are measurable on the ground. Not all enterprise goals and objectives will be applicable in all areas for team-based scorecards.

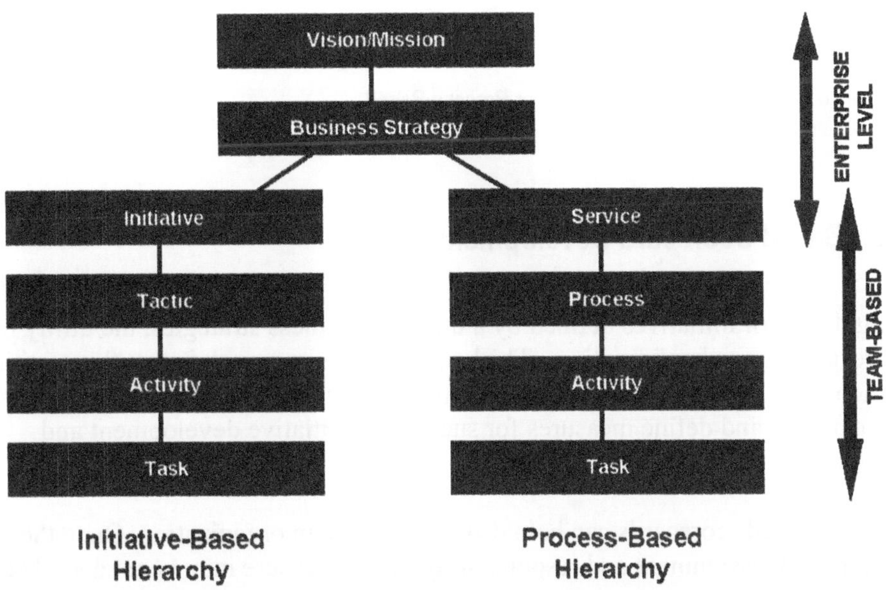

Figure 4-2. Team-Based Scorecards Can Be Based on an Initiative or Process Hierarchy.

While the roll-up to the enterprise level may in some cases obscure the variability in different parts of an organization's processes or initiatives (e.g., the averaging effect), more richly reported data (e.g., average with minimum/maximum deviations) can still show individual area differences.

A utility needs to select the best approach (initiative-based or process-based) or a hybrid approach to rolling out the scorecards organization-wide as shown in Figure 4-3. Individuals or organization units should be able to identify and have responsibility for a set of initiative-based or process-based measures that may come from multiple scorecards. This handful of individual or organization unit measures must be clearly relevant to their work as well as aligned with achievement of goals and objectives on the scorecards.

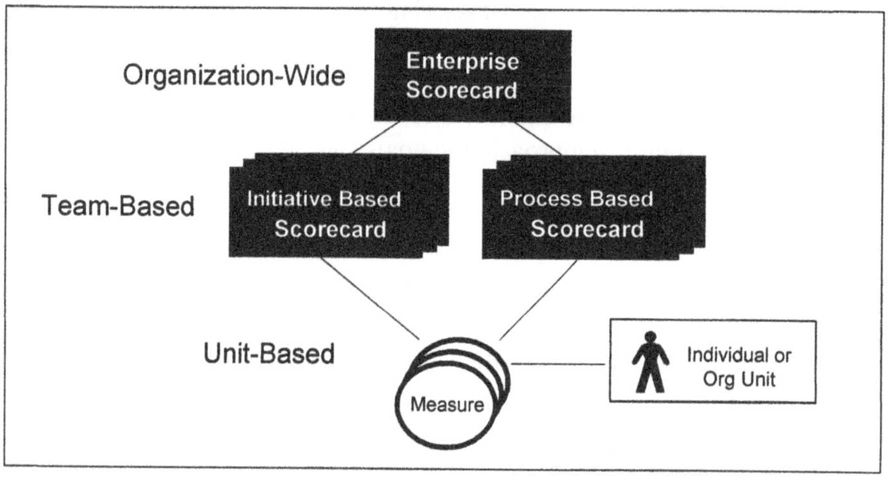

Figure 4-3. Balanced Scorecards Can Be Initiative- or Process-Based.

4.2 Team-Based Scorecard Development

Using the set of initiatives defined by a utility's business strategies, the utility may develop scorecards for its key initiatives. The benefit of this approach is that these initiatives are intended to achieve a desired change in the organization. Initiative teams—usually cross-functional—can own and define measures for successful initiative development and implementation.

Initiative-based scorecards can help drive change in an organization. Once the program or project is completed, assignment and responsibility of the measure may be unclear. Measures may not be distributed evenly across the organization or processes, dependent on the utility's initiatives and business strategies. However, these drawbacks should not discourage this approach as the normal evolution of scorecards and measures can deal with these changes once the measures are established and in use.

Process-based scorecards may be developed for business processes, particularly core processes identified in the business process framework. This approach may transcend organizational structures to measure performance of the complete effort of providing a particular service such as wastewater collection and treatment. Since measures and performance are managed over the long term, related communication to employees can be reinforced to keep goals and performance clear.

If a utility is looking for substantial change and has just embarked on a new strategic direction, initiative scorecards may be developed for each business strategy. If a utility is embracing continuous improvement of the business processes, a process scorecard approach may fit the best. Some examples of both initiative- and process-based scorecards are shown in Table 4-1. Again, a utility may select either approach or a hybrid of both key processes and key initiatives.

Table 4-1. Example of Initiative and Process Scorecards.

Initiative Scorecards	Process Scorecards
• CMOM/SSO Program	• Water Treatment and Distribution
• Asset Management Initiative	• Wastewater Collection and Treatment
• Security Initiative	• Solids Handling and Disposal/Reuse
• Customer Outreach Program	• Procurement Services
• Financial Stability Initiative	• IT Services
• Succession Management Initiative	• HR Services
• Energy Management Program	• Finance and Accounting Services
• Managed Maintenance Program	• Fleet Management Services
• Employee Accountability Initiative	• Engineering and Construction Services

A third and obvious team-based scorecard approach is founded on organizational structure. While possible to create organizational unit-based scorecards to further the rollout of measures, if this approach is used initially, it can reinforce organizational silos. It sometimes creates competing measures across an organization. In contrast, process- or initiative-based scorecards tend to breakdown the organizational silos and form the basis for common objectives and measures. Organizational unit scorecards are better used after the initial development of initiative- or process-based scorecards.

An alternative to the initial formation of organizational unit-based measures is the mapping of measures from process- or initiative-based scorecards as shown also in Figure 4-3. This mapping process is useful to determine the extent of organizational coverage after the enterprise and process- or initiative-based scorecards have been developed. The same mapping can be done at an individual employee level to determine what measures should drive individual performance and team-based performance. As another check and balance, an employee should be able to focus on several of the most relevant measures (e.g., five to 10 measures) based on reviewing scorecards applicability to his or her own work and responsibilities.

As shown in Figure 4-4, development and implementation of team-based scorecards follows that of the enterprise scorecard. Following the seven-step process from Chapter 3.0, the first three steps must be completed initially at the enterprise level before the team-based scorecards should proceed. This allows for measures to be the same or aligned between the team-based and enterprise scorecards. Then each team-based scorecard can continue with implementation that may add additional measures unique to that process or initiative that do not appear on the enterprise scorecard. In addition, not necessarily all measures from the enterprise scorecard will be applicable to each process or initiative.

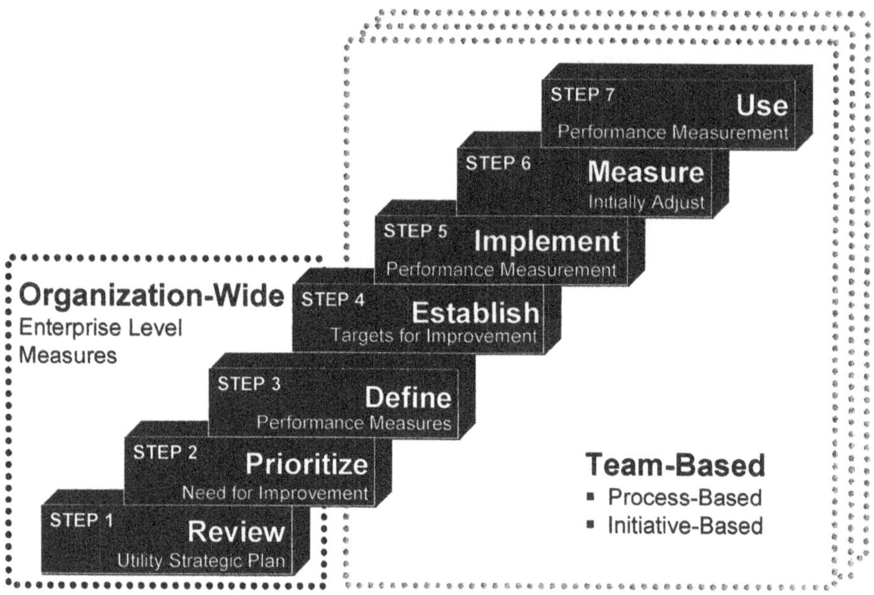

Figure 4-4. Team-Based Implementation Develops Measures Meaningful for Day-to-Day Use.

An alternative approach that follows the same seven-step process is for Steps 3 and 4 to start with the team-based measures, determine which ones apply organization-wide, and then define the enterprise measures and targets. In many cases, both top-down and bottom-up approaches are needed to get good alignment in enterprise and team-based scorecards.

The overall development, implementation, and rollout of enterprise and team-based scorecards can be accomplished over several months. As shown in Figure 4-5, completing team-based pilots that show how the implementation can be accomplished in some selected areas allows for piloting the entire system in 6 to 9 months. The remaining team-based scorecards can be prioritized for later rollout over a longer period (e.g., 12–18 months). The duration of the rollout will depend on the size and complexity of the organization as well as the urgency and commitment of the utility leadership to make it happen.

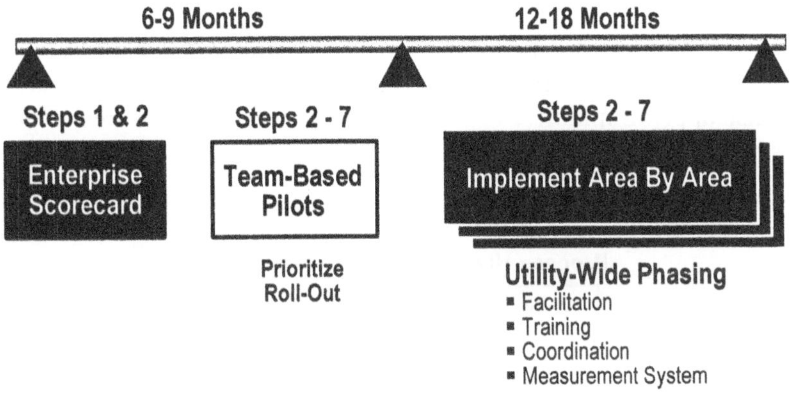

Figure 4-5. Implement Team-Based Scorecards in Phases Following Initial Pilots.

4.3 Measurement Aggregation or Roll-Up

The method for aggregation or roll-up of measures in process-based scorecards is shown in Figure 4-6. As shown in this hierarchy and chain of processes, efficiency measures can be summed or averaged as a technique to roll-up to the enterprise level. For instance, each process in a chain (e.g., water production, treatment, distribution) will contribute part of the overall cost of production (e.g., $/volume produced) that can be summed. In addition, costs of all the activities in a process, such as operations and maintenance activities, can be summed.

For effectiveness or quality measures, a link or relationship such as the *so-that* chain presented in Chapter 3.0, is appropriate for roll-up purposes. The same aggregation or roll-up techniques may also apply to initiative or organization unit-based measures, depending on related business processes.

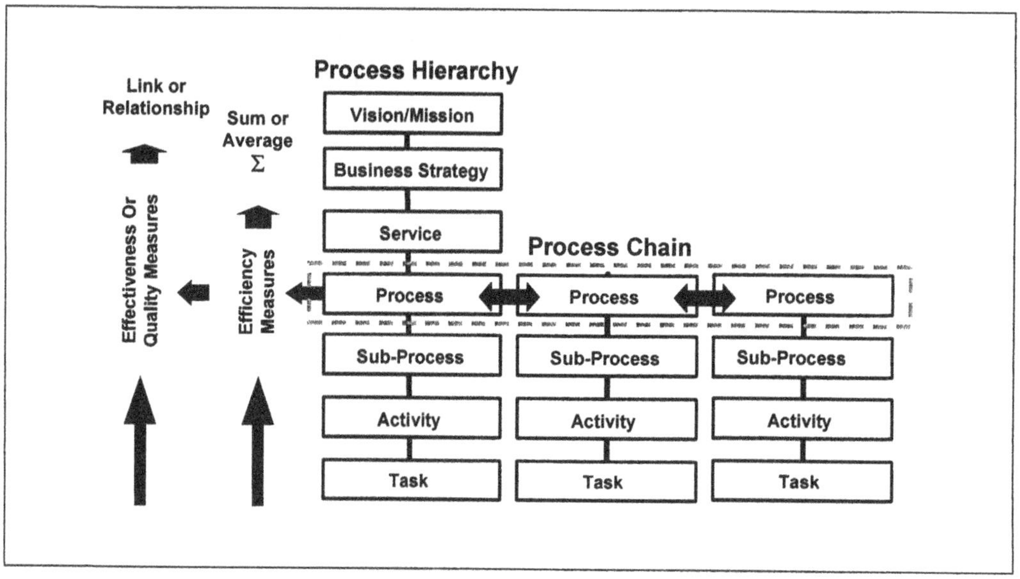

Figure 4-6. Performance Measures Will Aggregate Using Process Chains or Hierarchy.

4.4 Criteria for Good Measures

Additional lessons from the pilot projects and other leading practices include having simple guidance on what makes a good measure. Below are some examples of useful and not useful measures:

Useful Measures:

- Drive performance improvements (desired behaviors)
- Actionable and relevant to work activities (even if not entirely "within my control")
- Can be updated on a regular frequency (data collection is available or possible)
- Clearly defined with targets (units, how measured, baseline or trend available)
- Directly tie to objectives and goals on the scorecard
- Are process-based or initiative-based (organization or individual ownership follows)

- Use industry comparable or standard measures where fitting (e.g., QualServe measures)

Not Useful Measures:
- Have little impact on related actual work activities
- May be consolidated too much, making them less meaningful
- Too complex to understand or too complex to measure
- Impossible to collect data or more trouble than it's worth
- Just because it's easy to measure
- Occurs too infrequently to measure progress and trends
- Poorly defined and inconsistent in how measured
- Viewed as not actionable, unable to influence, or entirely "out of my control"
- May be subject to manipulation just to look good

4.5 Improving the Seven-Step Process

Feedback from pilot utilities on implementing the seven-step process has reiterated some key points previously included as best practices (Chapter 2.0). As shown in Figure 4-7, both education with reinforcement and communication with commitment are necessary for successful implementation. These practices are different sides of the same coin, for education leads to the understanding necessary for embracing measures to improve performance, while continuous communication is part of the feedback loop in reinforcing the system. Leadership from the top of the organization with commitment at all management levels is necessary for performance measurement to become part of a utility's culture.

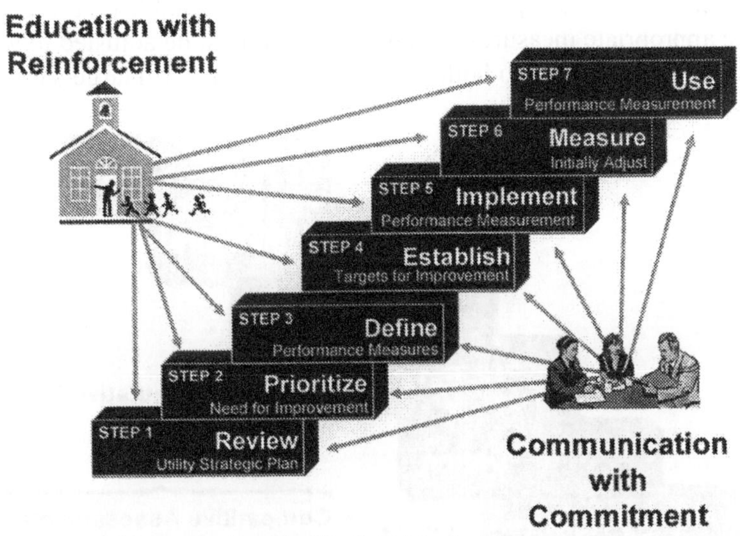

Figure 4-7. Education and Communication Must Be Built into Each Step.

Additional insight on specific steps in the process is noted here. These are expansions to the descriptions of each step as previously explained in Chapter 3.0. As shown in Figure 4-8,

Steps 1 and 3 can create reinforcement between the strategic plan and performance measures. From the strategic plan development or update, there is need for new measures to drive the organization (top down). From development of measures for strategy goals and objectives, the plan can be further clarified and focused (bottom up).

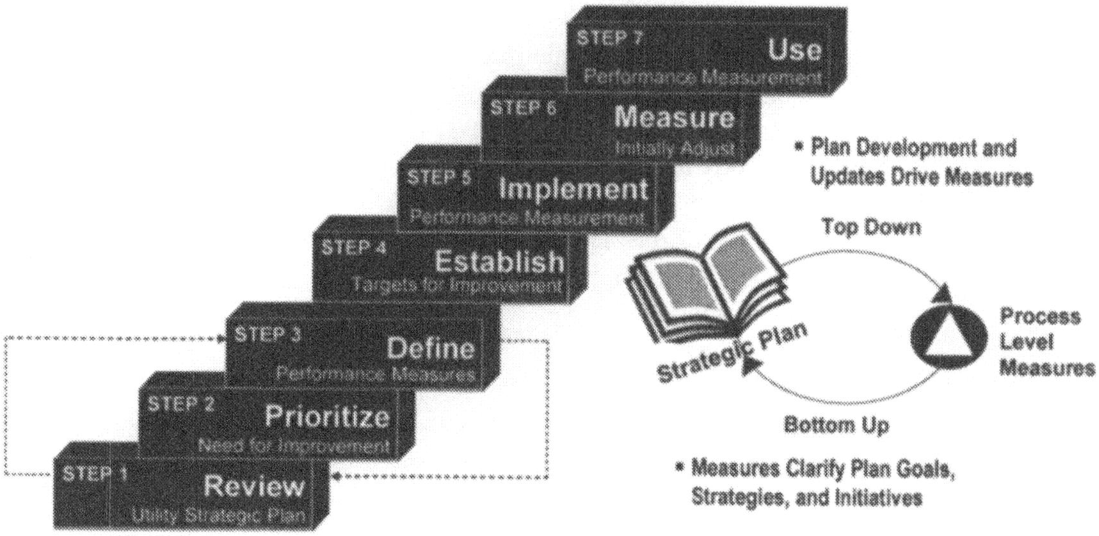

Figure 4-8. Development of Scorecard and Measures Must Align with Utility Strategic Business Plan.

As shown in Figure 4-9, Steps 2 and 4 are related in using the results of competitive assessments or reviews of current performance (e.g., QualServe or benchmarking) to set targets for improvement. Many times these studies will quantify or qualify a gap which is the opportunity for improvement. These gaps can become the basis to establish new targets for performance using the appropriate measures. Often, the target must be adjusted to what is practical for the period of performance so both long-term and short-term targets can be set.

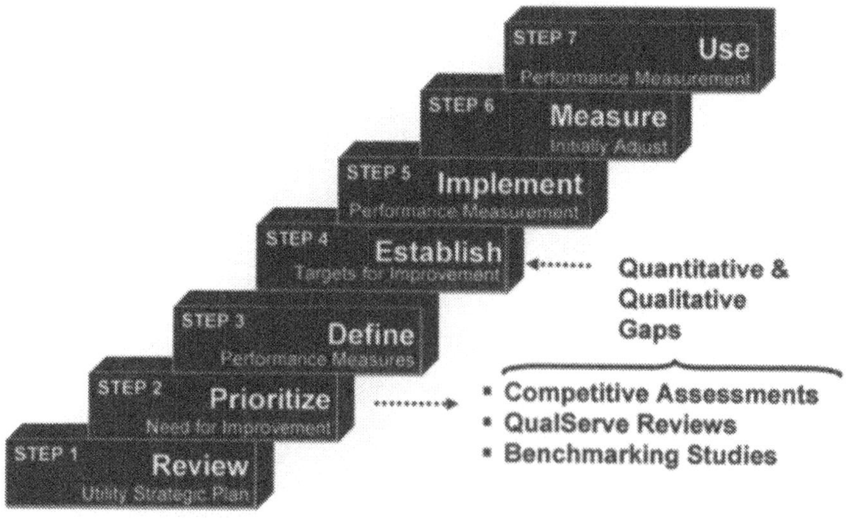

Figure 4-9. Improvement Drivers Can Be Used to Prioritize Needs and Set Targets.

As shown in Figure 4-10, to put the performance measures into regular use requires implementing and initially adjusting the measures. In order to implement, the roles and responsibilities for those involved in collecting, reporting, reviewing, and acting on the measurement results must be clear. Supporting technology is typically necessary to collect and store the data as well as present it in a useful way (as described later in this chapter). In addition, supporting human resource systems, such as team-based incentives and employee development plans, can reinforce performance measures.

Figure 4-10. Using Measures to Drive Performance Requires Systematic Data Collection, Tracking, and Actions to Get Results.

4.6 Collecting the Measures

In the implementation steps of the process (Steps 5–7), a number of additional items must be defined for each measure. Some of the key items include:

- Method of Measurement—includes appropriate description of data elements, calculations, and what data must be included or excluded so that the measure can be both accurate and repeatable.
- Frequency of Measure—a number of related frequency issues include:
  - Frequency of data collection or sampling
  - Frequency of recording (sometimes multiple samples with data manipulation)
  - Frequency of reporting (sometimes with data manipulation, aggregation and calculations)
  - Frequency of communicating for appropriate actions
- Period of Target—a number of related periods for the target include:
  - Period for collection of baseline data prior to target setting (if needed, i.e., no baseline data exists)

- Period over which the measure will be compared to the target (e.g., quarterly, annually)
- Period for target achievement related to any incentives (e.g., annual or fiscal year)
- Period for adjusting or re-setting the target

- Sources of Data—includes the information systems technology (described later in this chapter) as well as any existing databases or surveys (including ongoing periodic surveys) and human data sources in cases where people make observations and record data from other sources.
- Responsibilities for the Measures—a number of related responsibilities include:
  - Responsibility for data collection
  - Responsibility for analysis and verification
  - Responsibility for reporting and communication
  - Responsibility for taking action based on understanding and dialogue with parties involved
  - Responsibility for refinement or periodic updating of measures and/or targets

4.7 Considerations for Technology Implementation

Once the desired performance measurements are defined, a key challenge will be to determine the data requirements, location of data, and how it can be captured and loaded into the performance measurement system. Often the same data are called different names by different parts of and organization. An important step is to make sure that the data the utility thinks it is getting is the data that the utility is in fact getting. Given the overlap between data used by the existing applications (SCADA, WMS, LIMS, etc.), and data needed for the performance measurement system, it seems clear that there is no one-size-fits-all technology solution for the performance measurement system. Each utility must examine its own technology infrastructure and define a solution most fitting for their environment.

There are a number of functions and features that are desirable in any technology solution for collecting, manipulating, tracking, and reporting performance measures. The functions break down into three major categories: data input, data manipulation, and data presentation. Each of these categories is discussed below.

1. *Data input:* There are several considerations related to data input as follows.
   - *Level and nature of automation:* It is likely that some of the data can be automatically imported into the performance measurement technology system. The level and type of automation will vary significantly. This will be a function both of the measure itself (how often does it make sense to collect it) as well as a function of the computer system providing the data (how often can the system provide updates). In addition, there will be some measures for which data must be manually input. Regardless of the method used, it is critical that information about the data source be identified and maintained. By capturing this information, it is possible to track various pieces of data back to their sources when questions arise.

- *Timing:* Decisions must be made for each data source (and perhaps each datum) regarding the timing of insertion into the performance measurement system. Effectively interpreting and analyzing data requires a good understanding of the timeliness of the data and the time periods covered by the data.
- *Data format and validation:* In this case, data format includes two key issues: both the specific format as well as the unit of measure represented by the datum. While the unit of measure might be implicitly understood (everybody knows this means weekly work orders) in the source computer application, this will not be true when the data is imported into another computer system. Additionally, the format of the data must be handled. Last, it is important that any automatic data importing include validation of the data to reduce the risk of incorrect data corrupting the performance measurement system.
- *Data granularity:* The utility must anticipate the needs of all employees using performance data and how the data is to be manipulated or presented. For example, a performance measure for the organization may be cost per unit volume of sewage treated. When analyzing why the unit cost is different than targeted, a person may want to see the cost contributions of liquids and solids treatment separately or even finer breakdown of costs. The amount of drill-down desirable or practical depends on the level and nature of automation and the extent to which performance measurement is done within the organization. A rule of thumb is to include within the performance measurement system one level of drill-down below the lowest organizational level for which measures are defined. Further drill-down would be possible by using another system, such as a CMMS.

2. *Data manipulation:* People will want to view the data in the performance measurement system in a variety of ways; particularly, as more data is in the system and their data analysis needs change, they will want to see different views. This manipulation includes selecting different subsets of data for viewing and selecting different time ranges. They may want to select some data for further analysis. Drilling down into deeper levels of detail may be desirable.

3. *Data presentation:* Data presentation is closely related to data manipulation. As people view data and analyze it, they want to manipulate it depending on the results of their analysis. Effective data presentation is crucial to an active use of information. There are a number of aspects to data presentation, including these items:

   - *Information display:* It is nice to provide the information in a variety of methods, including text, charts (bar, pie, scatter, etc.), and to provide the ability to select the display option and configuration (the method of display). Users desire control over formatting, including use of color and fonts.
   - *Preselected views:* The information desired varies by role. The presentation of information according to the user's role, potentially remembering previous configurations of that user, becomes desirable. It is useful to provide a predetermined set of information by role or position and to allow this predetermined view to be customized. This is often called a dashboard display,

because the predetermined view provides summary critical information at a glance, much like the dashboard on a car.

In addition to the functions and features identified above, there are some general considerations that apply to any technology solutions. These include the following:

- *Ease-of-use:* Any computer application must be easy to use. A complex and difficult-to-use application will cause great resistance among the very population it is intended to serve.
- *Scalability:* The need for scalability varies depending on the organization's environment. Scalability is the ability of a solution to scale up, that is, to handle an increased number of users or transactions without dramatically requiring more computing or labor resources.
- *Robustness:* The application must work well and handle inevitable errors with grace. A fragile application that breaks frequently will not stay in use very long.
- *Access:* The application must be easily accessed by people for whom it is intended. If people must make a special effort to login to the application or to find a particular computer that runs the application, it is less likely to be used.

As indicated above, automating a performance measurement system requires integrating the system with other computer applications. The more integrated systems become, the more interdependent they become. Changes in a source system often require changes in a downstream system. Therefore, integration considerations must include long-term support aspects. For example, upgrades from one version of software to another version of software often include changing the internal data formats. These must be reconciled when systems are upgraded.

Currently, most utility data is under the control of the utility. That is, most applications lie within the utility boundary and do not cross-organizational boundaries. However, this is unlikely to remain the case in the future. Organizations are beginning to engage in strategic alliances that often require shared computing systems and resources. Additionally, with the advent of electronic commerce and electronic business, these once-clear boundaries are shifting. For example, as the procurement process moves onto the Internet, the data location will also be moving. It will be necessary to address these changing relationships in the future. There are a number of formats used to import/export data (ASCII, EBCDIC, comma delimited, XML, etc.). The specific format and method used is not particularly important, so long as it is understood exactly what data are being transferred and the export/import routines are synchronized.

There is a wide diversity of possible supporting technology systems for a performance measurement system, ranging from simple spreadsheets to highly complex and sophisticated systems. Rather than having the technology drive needs, it is important to ensure that needs drive technology. Regardless of what solution is selected, it will evolve over time due to changing needs as well as changing technology. There is no clear market leader or single common approach to implementing a performance measurement system technology solution. This is a somewhat new area for vendors with products rapidly evolving and frequent change in vendors. The situation is unlikely to change for some time.

There are some very good applications available in the high-end range. These applications integrate well with any computer system, ranging from mainframe batch-oriented systems to real-time SCADA systems. As with any highly integrated solution, it is necessary to customize the application to specific systems.

For utilities just starting on a performance measurement system, the recommended strategy is to start with readily available tools such as spreadsheets and perhaps a simple database. Technology requirements must be determined over time. As measurements stabilize and requirements become clear, the utility can begin to look for a more comprehensive solution that links directly back to requirements. As the program encompasses more and more of the organization, it will become more important to select and implement an enterprise solution.

In general, there are six categories of software solutions available today. They are described briefly below:

- *Office Productivity Tools:* This includes the typical suite of office productivity tools such as spreadsheets, word processors, and databases. These tools are readily available, have a low entry cost, and provide a large base of readily-trained users. They allow a tremendous amount of flexibility and can be effective entry-level mechanisms. However, they do not scale well, both in terms of numbers of simultaneous users and with large amounts of data. The ease-of-use tends to be highly variable as it is largely dependent on the skill and perspective of the person creating the custom application. They do not integrate well with real-time data systems.
- *Enterprise Resource Planning (ERP) Solutions:* This category of software includes products from vendors such as SAP, PeopleSoft, Baan, and Oracle, as well as third party add-on packages. These applications are pre-integrated, thus reducing some of the data definition issues identified above. Depending on the extent to which they have been implemented at a utility, these applications provide a significant amount of the required data. Using the data they provide as a part of the performance measurement system will increase the benefit of installing these complex and inflexible systems. They require specialized and skilled computer professionals to extract and manipulate the desired data. Although some of these applications profess to provide performance measurement systems, they are limited to the specific modules installed. Thus, they are very limited in the extent to which they support areas such as process control and laboratories.
- *Data Warehouses/Enterprise Information Systems (EIS):* The function of these applications is to make analysis of operational data easy and to provide support for making decisions. They are designed with the ability to import data and typically have data transformation capabilities built-in. They are built on relational database management technology such as Oracle, SQL Server, or DB2. These custom systems are excellent at handling large amounts of data. They often include visual analytical tools to enhance easy analysis of the data. However, they require specialized knowledge to program and maintain.
- *Industry Applications (WMS, SCADA, CIS, GIS, LIMS, etc.):* Many of the applications used in utilities include some support for performance measures. In

combination with other tools, they can provide the basis for a very effective performance measurement system. The amount of integration varies from utility to utility and across applications. They often have different user interfaces in each application and generally do not have the same users in common (different people use the WMS vs. the SCADA system, etc.). To provide an enterprise-wide view requires extending these solutions. Of special note in this class of software are SCADA historians. These software applications are very good at collecting data and providing graphical views of that data. They have trending capabilities built in and are designed to handle real-time or near real-time data. Most historians have limited storage abilities ranging from several months to one year.

- *Enterprise Performance Measurement Applications:* These are high-end applications, specifically designed to provide effective performance measurement analysis, tracking, and reporting with standard views (e.g., CorVu, SAS, Panorama Business Views). They are designed to take data from other systems, transform the data, and provide it in a format useful for analysis. They provide easy-to-use graphical interfaces. Some of them handle real-time data on a summary roll-up basis. They have a high initial entry cost, but if used effectively, provide a cost-effective solution from a total life cycle perspective.

- *Enterprise Application Integration (EAI) Tools:* This category of software is somewhat different from those identified above. Rather than providing users with an environment for data analysis, they focus on connecting applications and sharing data between systems using pathways known as middleware. This enterprise-level software is often used as the glue for some of the other systems mentioned above.

These categories of software can be used in combination with each other; they are not exclusive solutions. In fact, most performance management solutions will be a combination of two or more of these.

CHAPTER 5.0

ADDITIONAL RESEARCH AND UTILITY NEEDS

5.1 Research Needs Overview

The need for additional research presented in this chapter is the result of issues and ideas from both this project as well as the author's experience with other utilities engaged in performance measurement. There is no attempt to prioritize these research needs as every utility has its own priorities.

5.2 Use of Prescribed Water/Wastewater Industry Performance Indicators

Many utilities believe that measuring performance means adopting a set of prescribed measures regardless of their own unique situations. Additional research could show the results of utility performance by comparing those that adopt the prescribed measures with no additional system of measures versus those with systems such as BSC or strategy-based measures from a strategic plan. Additional comparison of results could be utilities that adopt prescribed measures versus those with measurement systems that do not use prescribed measures.

A sampling of case studies or a practice-based consortium study of a group of utilities could discover the best practices based on their real-world experiences using performance measures.

5.3 Education and Simple Tools

The pilots have shown that steps to developing performance measures are initially more important than the measurement system technology that is eventually implemented. With additional tools (spreadsheets and guidance documents) and training via a workshop, utilities can more easily implement performance measurement.

A set of tools to assist in developing and implementing performance measures would include templates and forms for developing measures with examples of typical measures as well as spreadsheets for collecting and reporting measures on a periodic basis. Without these tools, a utility embarking on a performance measurement program would need to create their own homegrown system or purchase commercially available software (typically priced from $50,000 to $100,000 or more depending on functionality and scalability). These tools should be simple, straightforward document templates and spreadsheets with examples of performance measures for typical water and wastewater utilities.

An orientation and training workshop could guide utilities in developing their own utility scorecard with interactive exercises. Participants can become familiar with the seven-step implementation process and use of the electronic tools described above to assist in scorecard development, implementation, and measurement tracking.

5.4 Linking Performance Measures to Incentives

An industry best practice for performance measurement is the linking of measures to incentives. While many public agencies have difficulty implementing monetary-based incentive systems, other ways for measures to be linked to incentives, including organizational, team-based, or individual, need to be further explored and tested. Both monetary and nonmonetary incentives should be researched for their affect on behaviors of individuals, teams, and the entire organization.

A sampling of case studies or a practice-based consortium study of a group of utilities could discover best practices based on their real-world experiences using incentives tied to measures.

5.5 Proof of Benefits in Using Performance Measures

The pilot projects demonstrated how to develop and implement a performance measurement system. They did not intend to capture or show the benefits of this effort beyond the initial gains from the development process itself. Case studies of the long-term tangible benefits of using performance measures exist in many businesses and are documented in references cited in this report. However, for the wastewater/water industry, there is only anecdotal evidence of these long-term tangible benefits

Additional research could be conducted to verify these benefits at wastewater utilities through case studies or other means.

APPENDIX A

GLOSSARY

Activities	The major components of the work done in a process. Each activity consists of input-process-output.
Balanced Scorecard (BSC)	A framework of performance measures from four perspectives (financial, customer, internal business process, and learning and growth) that are driven by the organization's overall vision and strategy. Developed by Kaplan and Norton, 1996.
Baldrige Award	The Malcolm Baldrige National Quality Award was established in 1987 in order to promote broad-based understanding of the requirements for performance excellence and competitiveness improvements and to promote sharing of information on successful performance strategies.
Benchmarking	A structured approach for looking outside the organization by studying and adapting the best outside practices to complement internal operations and creative ideas. Benchmarking is a learning experience—observing what the competition is doing now and projecting what that performance will be in the future (Xerox's definition).
BSC	Balanced scorecard
Business Perspective	A view of performance from a specific standpoint such as the customer, investor, or regulator. Business perspectives for public utilities are customer, financial, internal business, and learning and growth.
Business Process Reengineering (BPR)	The rapid and radical redesign of strategic, value-added business processes—and the systems, policies, and organizational structures that support them—to optimize the work flows and productivity in an organization.
Business Service	Services that support and enable a utility's core services. These services may be provided internally or by outside service providers (i.e., outsourced).
Business Strategy	A plan for achieving a portion of the strategic direction.

Capital Improvements	Investments in a utility's buildings, structures, equipment, and other forms of capital plant.
Change	Alterations, transformations, and substitutions; something new starts or stops and takes place at a particular point in time.
Closed-Loop Process	A process in which an output serves as feedback to adjust an input to the process.
Competitive Gap	The value of the difference from a competitive measure, also called a competitive margin.
Competitiveness Assessment	A review of how well the utility performs against best practices of the most competitive utility organizations.
Competitiveness Goal or Target	Goal or target defined to be achieved in order to become or remain competitive.
Compound Loop Control	Control of a process variable based on an inner loop of control being adjusted by outer loop of control.
Constraint	A limitation on the mission and scope of the reengineering project identified by senior management.
Continuous Improvement	Improvement efforts in performance that may be radical as well as incremental.
Controls	Factors manipulated to achieve results (e.g., contact time, workforce training, design criteria).
Core Competencies	Abilities required to deliver services, which integrate many skills, are difficult to master, provide competitive advantage, and provide substantial and unique value to the customer.
Core Process	All the functions and sequences of activities (regardless of where they reside in the organization), policies and procedures, and supporting systems required to meet a marketplace need through a specific strategy. It includes all functions involved in the development, production, and provision of specific services to particular customers.
Core Service	Services offered to customers, which meet a company's mission. Core services define the business of a company.

Cost — The non-value-added activities and resources incurred in meeting the requirements of the customer.

Cost/Benefit Analysis — A quantitative assessment relating project benefits to the costs necessary to bring about the benefits. Often a critical tool in justifying the full-scale implementation of the design.

Design — The phase in reengineering for rethinking and designing the best way to accomplish work. The design phase includes the development of detailed business maps, procedures, and operating policies and the design of supporting infrastructures (e.g., performance measures, job and organizational design, compensation systems, etc.).

Detailed Design — A map and related documentation depicting how work will get done in the future.

Effectiveness — Doing the right things. The degree to which performance objectives are achieved and can be considered reflections of the quality of the unit's performance. Response times are common effectiveness measures, and a specific example in the wastewater industry is the percentage of repaired sewer pipe still functioning properly after 10 years time.

Efficiency — The resources used or cost per unit of output. These measures provide information about the production of an output at a given level of resource use and demonstrate an entity's relative efficiency. The term *productivity indicator* is sometimes used instead of *efficiency indicator*.

Enterprise Balanced Scorecard — A scorecard for the organization as a whole, representing a high-level report of performance.

Explanatory Factors — Factors that affect performance, but are outside the ability to control. (e.g., population served, area of service, rainfall)

Feedback — Process of adjustment based on the measured outputs.

Feedforward — Process of adjustment based on the measured inputs.

GASB	Government Accounting Standards Board. This board's mission is to establish standards of accounting and financial reporting for state and local governments.
Implementation	Full-scale rollout to the new design, including the execution of the performance management components.
Inputs	The monetary and non-monetary resources expended (often expressed in dollars) in providing a product or service.
Instrument	To implement measures of performance and/or status, i.e., to install instruments of measure.
Intangible Benefit	A benefit that is difficult to quantify yet critical when evaluating potential investments.
ISO	International Standards Organization. This organization has established standards for quality management (ISO 900x) and for environmental management (ISO 14000).
Leading Practice	Practices of world class organizations. Also called best practice.
Measurement	The dimension, quantity, or capacity of a process; see Results Metrics and Process Metrics.
Methodology	A body of methods, procedures, and working rules.
Metric Benchmarking	The quantitative measurement of performance in terms of inputs (i.e., labor, capital, land) outputs (i.e., water delivered, wastewater treated), outcomes (i.e., customer satisfaction, clean rivers, and lakes), and the relationships among them (i.e., costs per 1,000 gallons delivered or treated).
Model	A representation of the structure and operations of a business or other organization used to understand the process of the organization. A preliminary idealized structure used to test a design.
Motivators	A factor that can increase motivation (i.e., rewards, recognition, etc.).
Performance Measurement System	The process by which utilities continuously improve their business performance.

Performance Review Program	Any organized effort of determining and improving performance, including competitiveness assessments and benchmarking.
QualServe™	A performance review program developed by the American Water Works Association (Awwa) and the Water Environment Federation (WEF) for water and wastewater utilities. It includes components of self-assessment, peer review, and the benchmarking clearinghouse.
Strategic Direction	A statement of purpose composed of the organization's mission, vision, and values.
Strategic Initiative	An operational initiative that directly carries out a high-level business strategy.
Strategic Plan	A strategic plan states strategic direction of the organization and defines business strategies and operational initiatives to achieve the direction.
Tactical Initiative	An operational initiative that directly carries out a low-level business strategy.

APPENDIX B

BIBLIOGRAPHY AND REFERENCES

American Productivity and Quality Center (APQC), International Benchmarking Clearinghouse (2002) *Performance Measurement: Implementing the Balanced Scorecard.* Houston, TX.

American Productivity and Quality Center (APQC), International Benchmarking Clearinghouse (1999) *Achieving Organizational Excellence Through the Performance Measurement System.* Houston, TX.

American Productivity and Quality Center (APQC), International Benchmarking Clearinghouse (1998) *Enterprisewide Systems: Consortium Benchmarking Study Best-Practice Report.* Houston, TX.

American Productivity and Quality Center (APQC), International Benchmarking Clearinghouse (1997a) *Benchmarking Benchmarking: Shared Learnings for Excellence*. Houston, TX.

American Productivity and Quality Center (APQC), International Benchmarking Clearinghouse (1997b) *Enabling Success in Corporate Performance Measurement: Consortium Benchmarking Study Best-Practice Report*. Houston, TX.

American Productivity and Quality Center (APQC), International Benchmarking Clearinghouse (1996) *Corporate Performance Measures: Consortium Benchmarking Study Best-Practice Report*. Houston, TX.

American Productivity and Quality Center (APQC), International Benchmarking Clearinghouse (1992) *Process Identification Framework.* Houston, TX.

American Water Works Association (AWWA), MainStream (1999) *Charlotte Used Scorecard to Hone Performance*. Denver, CO.

Awwa Research Foundation (2004) *Definition of Performance Indicators for Water and Wastewater Utilities.* Denver, CO.

Awwa Research Foundation (AwwaRF) (2002) *QualServe Performance Indicators for Water and Wastewater Utilities*. Denver, CO.

Awwa Research Foundation (AwwaRF) (2001) *Consortium Benchmarking Methodology Guide.* Denver, CO.

Awwa Research Foundation (AwwaRF) (1997) *The Utility Business Architecture: Designing for Change*. Denver, CO.

Awwa Research Foundation (AwwaRF) (1997) *Guidance for Self-Assessment and Peer Review for the AWWA QualServe™ Program*. Denver, CO.

Awwa Research Foundation (AwwaRF) (1996) *Performance Benchmarking for Water Utilities*. Denver, CO.

Ammons, David N. (1996) *Municipal Benchmarks: Assessing Local Performance and Establishing Community Standards*. Thousand Oaks, CA: Sage Publications.

Association of Metropolitan Sewerage Agencies (AMSA) and Association of Metropolitan Water Agencies (AMWA) (1998) *Thinking, Getting, and Staying Competitive: A Public Sector Handbook.* Washington, D.C.

Camp, Robert C. (1989) *Benchmarking: The Search for Industry Best Practices that Lead to Superior Performance*. Milwaukee, WI: ASQC Quality Press.

Colwell, W.L. and Koletar, J.W. (1984) Performance Measurement for Criminal Justice: The Federal Bureau of Investigation Experience. *Public Productivity Review* 3: 207-224.

Deming, W. Edwards (1993) *Out of The Crisis*. Boston, MA: Massachusetts Institute of Technology, Center for Advanced Engineering Study.

Deming, W. Edwards (1993) *The New Economics for Industry, Government, Education*. Boston, MA: Massachusetts Institute of Technology, Center for Advanced Engineering Study.

Epstein, Paul D. (1992) Measuring the Performance of Public Service. *Public Productivity Handbook*. New York, New York: Marcel Dekker, Inc., 161-194.

Epstein, Paul D. (1988) *Using Performance Measurement in Local Government: A Guide to Improving Decisions, Performance and Accountability*. Denver, CO: National Civic League Press.

Fischer, R.J. (1994) An Overview of Performance Measurement. *Public Management* 76: S2-S8.

Glover, Mark (1992) *A Practical Guide for Measuring Program Efficiency and Effectiveness in Local Government*. Tampa, FL: Innovation Group Publications.

Governmental Accounting Standards Board Research Report (1990) *Service Efforts and*

Accomplishments Reporting: Its Time Has Come. Washington, D.C.

Harrington, H.J. (1991) *Business Process Improvement: The Breakthrough Strategy for Total Quality, Productivity, and Competitiveness*. Milwaukee, WI: ASQC Quality Press.

"*Harvard Business Review on Measuring Corporate Performance*." (1998) Boston, MA: Harvard Business School Press.

Harvey, J. (1987) Measuring Productivity in Professional Services. *Public Productivity Review* 44: 29-38.

Hatry, P. and Fisk, D. (1992) Measuring Productivity in the Public Sector. *Public Productivity Handbook*. New York, NY: Marcel Dekker, Inc., 139-160.

Holzer, M. and Halachmi, A. (1996) Measurement as A Means of Accountability. *International Journal of Public Administration* 11,12,19: 1921-1944.

International City/County Management Association (1997) *Comparative Performance Measurement Data Report*. Washington, D.C.

International City/County Management Association (1995) *Accountability for Performance Measurement and Monitoring in Local Government*. Washington, D.C.

International City/County Management Association (1992) *How Effective Are Your Community Services? Procedures for Measuring Their Quality*. Washington, D.C.

International Water Association (IWA), (2000) *Performance Indicators for Water Supply Services*. IWA Publishing, London, UK (as part of its Manuals for Best Practices Series).

Kaplan, Robert S. and Norton, D.P. (1996) *The Balanced Scorecard*. Boston, MA: Harvard Business School Press.

Kaplan, Robert S. and Cooper, R. (1997) *Cost & Effect: Using Integrated Cost Systems to Drive Profitability and Performance*. Boston, MA: Harvard Business School Press.

Lynch, Robert F. and Werner, T.J. (1994) *Reengineering Business Processes and People Systems*.

Managing for Results: Performance Measures in Government (1993). Conference Proceedings. Austin, TX.

Marshall, Wray, Epstein, and Frifel (1999) *21st Century Community Focus: Better Results by Linking Citizens, Government, and Performance Measures*.

McGowan, R.P. and Poister, T.H. (1985) Impact of Productivity Measurement Systems on Municipal Performance, *Policy Studies Review.* 4: 532-540.

Monahan, K.E. (2000) *Balanced Measures for Strategic Planning (A Public Sector Handbook).* Vienna, VA: Management Concepts.

Olve, Nils-Goran, Roy, J., and Wetter, M. (1999) *Performance Drivers: A Practical Guide to Using the Balanced Scorecard.* New York, NY: JohnWiley and Sons, Inc.

Paralez, Linda L. (1999) *Utility Benchmarking on the West Coast.* Journal of American Water Works Association. Denver, CO.

Performance Measurement Reports Research Packet. Tampa, FL: The Innovations Group Publications.

Player, S. and Keys, D.E. (1999) *Activity-Based Management: Arthur Andersen's Lessons from the ABM Battlefield* (2nd ed.) New York, NY: John Wiley and Sons, Inc.

Porter, Michael E. (1998) *Competitive Advantage: Creating and Sustaining Superior Performance.* New York, NY: Simon and Schuster Trade Publications.

QualServe Benchmarking Clearinghouse (AWWA/WEF) Performance Indicators (2002) *Pilot Survey Results* (Subscriber Members Report).

Senge, Peter M. (1990) *The Fifth Discipline, The Art & Practice of The Learning Organization.* New York, NY: Doubleday and Company, Inc.

Senge, P.M., Kleiner, A., Roberts, C., Ross, R.B., and Smith, B.J. (1990) *The Fifth Discipline Fieldbook, Strategies and Tools for Building a Learning Organization.* New York, NY: Doubleday and Company, Inc.

Spendolini, Michael. (1992) *The Benchmarking Book.* New York, NY: American Management Society/AMACOM Publishing.

Streib, Gregory and Poister, T.H. (1998) Performance Measurement in Municipal Governments *Municipal Year Book.*

Swiss, James E. (1991) *Public Management Systems: Monitoring and Managing Government Performance.* New York, NY: Simon and Schuster Trade Publications.

Tapscott, Don and Caston, A. (1993) *Paradigm Shift: The New Promise of Information Technology.* New York, NY: McGraw-Hill Company.

Taylor, David A. (1995) *Business Engineering with Object Technology*. New York, NY: John Wiley and Sons, Inc.

The U.S. Foundation for Performance Measurement. Washington, D.C.

Water Environment Research Foundation. (2000) *Developing and Implementing a Performance Measurement System: Volume I,* published by WERF, Alexandria, VA.

Water Environment Research Foundation. (1999) *Improving Wastewater Treatment Plant Operations Efficiency and Effectiveness.* Water Environment Research Foundation (WERF) Project Number 97-CTS-1. WERF, Alexandria, VA.

Water Environment Research Foundation. (1997) "*Benchmarking Wastewater Operations—Collection, Treatment, and Biosolids Management.*" Washington, D.C.

Water Services Association. (1994) *Waterfacts.*

APPENDIX C

PILOT UTILITY SUMMARIES

Four utilities piloted the seven-step process (Figure C-1) and related best practices in their own unique circumstances. Each pilot included some or all steps in the process. Results indicate that when the steps are carried out successfully, a useful performance-measurement system is established.

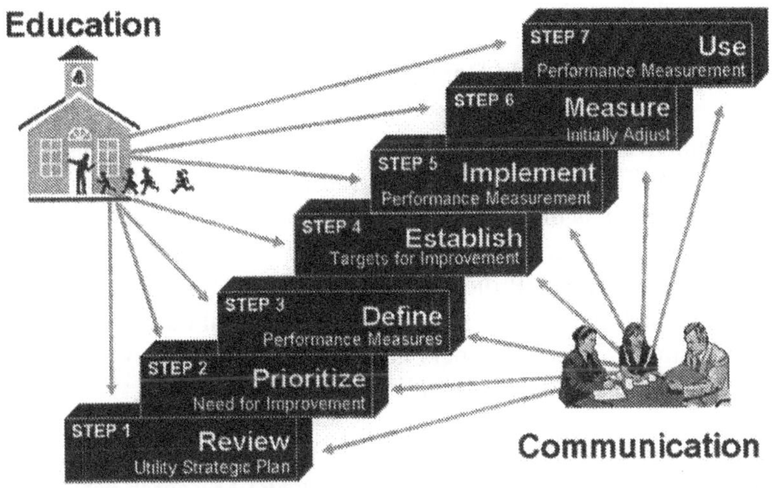

Summaries of each pilot implementation in terms of the individual utility's lessons learned and adjustments to the performance measurement framework, approach, or seven-step process is included below. The utilities and their individual focus areas were:

- City of Phoenix Water Services Department—A utility-wide scorecard was developed and linked to an updated department strategic plan. Implementation of scorecard goals, objectives, and measures was performed for two focus areas: managing maintenance at Union Hills Water Treatment Plant; and prevention of sanitary sewer overflows for the zero tolerance regulatory compliance program.

- Union Sanitary District—A treatment and disposal (T&D) process scorecard focused work teams on establishing targets and measuring performance. A refined scorecard linked to process and team-based scorecards for an enterprise view of utility-wide performance.

- Seattle Public Utilities—An updated strategic business plan directly linked measures from operational work plans and strategic initiatives to strategic business plan goals.

Refinement of the existing organizational performance system focused on balancing the types of measures and implementing lines-of-business measures.

- Central Contra Costa Sanitary District—An initial plant operations scorecard was defined with identified key performance improvement areas. Additional development and implementation of the scorecard aligned with the District's updated strategic direction.

Two of the pilot utilities, Seattle Public Utilities and Union Sanitary District, have previously implemented organization-wide performance measurement systems. All pilot participants have previously used performance measures in some areas. This project demonstrated an approach that helps utilities begin to embrace or expand performance measurement.

C.1 City of Phoenix Water Services Department

The City of Phoenix Water Services Department (WSD) (http://www.phoenix.gov) services approximately 1.3 million people who are served through more than 330,000 accounts. The total service area for Phoenix is more than 530 square miles. About 96% of the water used by Phoenix comes from surface sources, the Salt, Verde, and Colorado Rivers; the remaining 4% comes from wells. The Central Arizona Project and the Salt River Project deliver surface water to the city. Five water treatment plants, with a combined capacity of about 600-million gallons per day, provide water treatment. The department also operates two wastewater treatment plants with a combined capacity of about 220-million gallons per day.

This large utility was selected to participate in the Water Environment Research Foundation (WERF) and Awwa Research Foundation (AwwaRF) research project. The focus of this project was to demonstrate the practical and technical elements of performance measurement system development. WSD had two objectives for this project:

- Develop a WSD scorecard by reviewing and updating WSD's strategic plan
- Develop and implement two pilot projects.

WSD developed a utility-wide scorecard and linked it to an updated strategic plan. Additionally, scorecards were developed for two process-based focus areas.

Task 1: Define Utility Strategic Plan
A management leadership team (MLT) was formed; this team was charted to align WSD's vision and mission with business strategies. The MLT developed vision, mission, and best-in-class statements to set the strategic direction of WSD. Development of the vision and mission statement facilitated identification of key business strategies.

Task 2: Define Need for Improvement
A list of tactical initiatives was developed and correlated to the key business strategy that they support. Goals and objectives to support each business strategy were identified and categorized into one of the four scorecard perspectives (e.g., customer, financial, internal

business process and learning and growth). This resulted in development of WSD's enterprise balanced scorecard.

Task 3: Define Performance Measures

Two pilot projects were selected: maintenance management/computerized maintenance management system and zero tolerance regulatory compliance that focused on sanitary sewer overflows. Each pilot project used the WSD goals and objectives as the starting point to develop a unique scorecard . Pilot project teams consisted of key users and supervisors from each pilot area.

Each team began with identifying business processes performed to support the pilot area. Next, each WSD goal and objective was reviewed to determine if performance of the business process impacted (directly, indirectly, or not at all) achieving the stated WSD goal and objective.

For all goals and objectives marked directly or indirectly applicable, the process that supported achieving the goal and objective was identified. Specific measures were identified for each of these goals and objectives.

Task 4: Establish Targets for Improvement

For each priority goal and objective (i.e., those with measures assigned), a target measure was identified. The method and frequency of measurement were also determined. Additionally, responsibility was assigned to a person or group to gather and track actual measures.

Task 5: Implement Performance Measurement System

The transition from developing a scorecard to implementation included an analysis of measures and methods of measurement to determine whether a new activity, documentation requirement, or procedural change was being introduced. The first step in implementation was to train staff on basic performance measures and any new operating procedures and/or documentation requirements.

Task 6: Measure and Initially Adjust

The pilot projects gathered actual results for a period of 4–6 months with monthly reviews of data gathered. During this 4-6 month period, the pilot team asked the following questions:

Are the measures the right measures?
Are the targets realistic?
Are process or technology changes needed to gather information?

Task 7: Use the Performance Measurement System

Transitioning from pilot to full-time use, the team began to ask the question:
If a target is not being met why not?

When assigning a root cause to targets that are not met to support historical trending and analysis, they asked:

Are there processes, technology, training changes that need to occur to meet the target?

What technology is needed to support developing a performance measure system, gathering data and maintaining the performance measures over time?

Education/Communication

An overview of the WERF/AwwaRF performance measurement project and education on performance measurement frameworks was provided at three points in the project:

- Initial meeting of management leadership team
- Initial meeting of pilot performance measurement development teams
- Meeting of pilot performance measurement implementation teams

Lessons Learned

- It is important to scope project adequately, starting broad, defining focus. A service area should be picked that the team members are familiar with and that can be improved by systematically measuring progress.
- Teams must have a strong sponsor, champion, or management buy-in.
- Cross-functional teams do work.
- Meeting attendance and participation is mandatory.
- Teams should get into developing scorecard quickly, moving from abstract from something concrete. The important thing is to get started.
- It is important to review and adjust, as necessary. It must be remembered that this is a continuous improvement project.
- Teams may need to bring subject matter experts in to help develop measures, determine how to measure, and to set targets.
- A database or means of tracking data electronically is required.
- Perform root cause and actions to prevent recurrence for measures outside of target

C.2 Union Sanitary District

Union Sanitary District (USD) (http://www.unionsanitary.dst.ca.us/) is an independent special district that provides wastewater collection, treatment, and disposal services to approximately 318,000 residents and businesses of the cities of Fremont, Newark, and Union City, in Southern Alameda County, California.

USD was selected to participate in the WERF and American Water Works Association Research Foundation (AwwaRF) research project. The focus of this project was to demonstrate the practical and technical elements of performance measurement system development. USD had the following objectives for this project:

- Refine USD's current scorecard.
- Pilot a process-level scorecard.
- Provide employee access to scorecards.

Task 1: Define Utility Strategic Plan

USD had developed a performance measure balanced scorecard in 1998. This original effort developed scorecards at the USD and team level for technical support/customer service, treatment and disposal, business services, and collection services.

The USD- and Team-level scorecards served as the starting point for this effort. An executive team was formed. They began refining USD's scorecard by reviewing previous strategic planning sessions and previously established BSC objectives and measures.

Task 2: Define Need for Improvement

The executive team set goals for improving the effectiveness of the existing BSCs. Problems identified during this effort included:

- Measures were not connected to employee's day-to-day work
- Measures didn't roll up from team scorecards

These problems were translated into changes in how performance measures were developed and linked to USD's strategic plan.

Task 3: Define Performance Measures

Task 2 activities resulted in development of more specific performance measures and focused on a theme that employees could understand. Separate scorecards were developed that focused on operational excellence and safety. Measures were developed that 'roll-up' to USD measures and the scorecard was communicated graphically.

Process level scorecards were used to move focus of staff to the process level and to integrate previous performance measurement and improvement efforts. A pilot project was used to develop the treatment and disposal process scorecard. The scorecard was linked to USD's strategic plan and replaced three team scorecards. The need for improvement was determined by performing a self-analysis of competitiveness using the AMSA assessment tool.

The scorecard was built using a cross-functional team of employees. This process emphasized core processes incorporating best practices where appropriate. The main focus was on improvement and alignment with USD's scorecard.

Task 4: Establish Targets for Improvement

Setting targets forced further refinement of measures. Team input and benchmarking data were used to establish targets.

Task 5: Implement Performance Measurement System

The treatment and disposal group had implementation responsibility. The scorecard measures and targets were communicated via spreadsheets and graphs to employees by this group. Summarizing and reporting of data was delegated to the office assistant for the treatment and disposal workgroup, who had skill in database development.

Task 6: Measure and Initially Adjust

Treatment and disposal began implementing the BSC.

Task 7: Use the Performance Measurement System

Challenges in using the system included getting employee ownership and really using the data to make decisions.

Education/Communication

- Management as well as staff must be educated on performance measurement concepts, with the goal of moving from just collecting and reporting data, to using the data to make decisions.
- Employees should be surveyed on performance measurement concepts and attitudes, in order to determine their level of understanding and the input used to improve the process.

Lessons Learned

- The scorecard may have numerous drafts and refinements. There's a need to balance continuous improvement with stable measures that are inn place long enough to show trends.
- Self-assessment creates greater buy-in for the need to improve.
- Staff was already familiar with the concept of performance measurement; there was greater employee trust about how the data would be used than when the concept was originally introduced.
- Useful team measures were incorporated into process scorecards.
- Working at the process level discouraged team silos.
- The process is lengthy and iterative.
- Resistance comes because of the additional workload.
- It is difficult to limit the number of measures. It is important to resist the urge to include a lot of feel-good measures and focus on areas that need improvement.
- While upper management support is critical, there is also a need for someone in the organization assigned to keep the performance measurement effort moving forward, coordinate data collection and reporting, and assist management and staff with implementation.
- With much patience and persistence, measuring and improving performance can become an accepted way of doing business.

C.3 Seattle Public Utilities

Seattle Public Utilities (SPU) (http://www.seattle.gov/util/services/) provides more than 1.3 million customers in King County with a reliable water supply and essential sewer, drainage, and solid waste services for the city of Seattle. To deliver these basic services, SPU relies on a system of pipes, reservoirs, and disposal and recycling stations.

SPU was selected to participate in the WERF and AwwaRF research project. The focus of this project is to demonstrate the practical and technical elements of performance measurement system development. SPU had the following objectives for this project:

- Define high-level (benchmark) measures
- Identify measurement gaps and adopt best practices
- Develop process-based (line of business) measures

Task 1: Define Utility Strategic Plan

Employees throughout SPU were involved in the development of the 1999–2000 Strategic Business Plan. This plan identified six strategic goals based on the principles of customer satisfaction, environmental stewardship, and responsiveness to citizen community and customer needs. Performance measures were also developed to link employees' work to the strategic goals.

The 2001–2004 Strategic Business Plan expanded on the foundational principles of the 1999–2000 Strategic Plan by adding one strategic goal, refining goal statements, and relating strategic initiatives to the goals and results of the strategic plan.

Task 2: Define Need for Improvement

SPU's goals as identified in the strategic plan were mapped to SPU's lines of business (LOBs—water, drainage/wastewater and solid waste) and to strategic focus areas within functional areas (finance and administration, customer service, engineering, resource management, and field operations).

Existing LOB performance measures were correlated to the appropriate BSC perspective (e.g., customer, financial, learning and growth, and internal business process).

Task 3: Define Performance Measures

Performance measures were applied to SPU LOBs and within functional branches. Workshops were held with each LOB to review current measures for the LOB, develop example measures for a service area within an LOB, and develop suggestions for going forward.

There were 287 measures identified for SPU, of which 118 were directly tied to SPU's LOBs. The remaining measures were independent of a specific LOB. General comments on these measures include:

- Further clarity and consistency was needed on how measures were described (purpose, methods, units, etc.).
- Measures were under-represented for the learning-and-growth and customer perspectives.
- Some LOB services do not have measures.
- Measures do not rollup or aggregate to LOB high-level measures.
- Current measures reflect view of 'what can I control?"
- Current measures are effectiveness or quality measures; there are almost no measures for efficiency.

- Most of these observations were typical of organizations that had initially implemented measures without a specific framework.

The consumption service area within the water LOB was selected as an area of further review. Measures were further developed for this service area based on a framework of efficiency, quality, and effectiveness.

Education/Communication

- Educate LOB teams on performance measurement concepts

Lessons Learned

- Developing measures requires knowledge of the business process, a lot of thought, and staff time and commitment.
- Unused or misused measures lessen credibility for the performance measures program.

C.4 Central Contra Costa Sanitary District (CCCSD)

Central Contra Costa Sanitary District (CCCSD) (http://www.centralsan.org) is an independent local utility with a publicly-elected board of directors who serve 420,470 residents in various east Bay Area communities. CCCSD builds, operates, and maintains facilities required to collect and treat up to 45-million gallons of wastewater each day.

CCCSD was selected to participate in the WERF and AwwaRF research project. The focus of this project was to demonstrate the practical and technical elements of performance measurement system development. CCCSD had an existing mission, vision, values, strategic plan, and an existing performance improvement initiative. CCCSD had the following objectives for this project:

- Develop and implement a plant operations scorecard (POS) that creates strong and direct links with the CCCSD's strategic business drives and initiatives.
- Apply the seven-step model in developing and implementing the plant operations scorecard.
- Build an integrated system that balances a family of measures and establishes targets.
- Educate staff about the theory, development, and application of performance measures.
- Systematically collect data that supports rollupsof the POS and make it available via CCCSD's Intranet.
- Maximize CCCSD's staff involvement in the project.

Overall, their goals were to improve quality of service, achieve alignment with organizational goals, and improve effectiveness and efficiency. Initial plant operations performance was defined with identified key performance improvement areas. With the updated 2001 Strategic Plan, additional development and implementation of the plant operations' scorecard was aligned and linked with the district's goals.

Task 1: Define Utility Strategic Plan

CCCSD's mission, vision, and core values were developed in 1998 as part of a work redesign project. This process analyzed and modified internal work processes through the perspectives of customer service, cost, and quality and work life. A strategic plan was developed at the same time through a joint board-management effort with review and input from employees. At this time, the strategic plan and strategic areas were not directly linked to CCCSD's core values or continuous improvement process although they were interrelated.

In 2000, the work redesign project was renamed continuous improvement, and a committee was formed to guide the process. The CCCSD management team updated the strategic plan and the mission, vision, and core values.

A performance measures task force was formed that included personnel from operations, maintenance, and pumping stations, managers, supervisors and front-line employees. This team was chartered with reviewing the BSC, modifying it from four perspectives (customer, financial, internal business process, learning and growth) into three to match CCCSD's continuous improvement perspectives (quality of work life, customer service, and cost-effective services). An additional focus of this task force was to map CCCSD strategic plan elements to the process-improvement method to demonstrate how the continuous improvement program and benchmarking fit into performance measurement.

In late 2000 and early 2001, CCCSD's managers reviewed and revised the mission, vision, and values establishing strategic goals for CCCSD. This meeting included a review of how to rollout the strategic plan to all CCCSD employees, what metrics should/could be used, and how the strategic goals and metrics could be included in each employee's personnel evaluation.

Task 2: Define Need for Improvement

Since CCCSD originally developed a performance measures program in 1998, they asked themselves, "If given the opportunity to start over, what would you do differently?" They decided they would show how organizational management and efficiency efforts fit into the scorecard and performance measures framework process.

They identified areas for performance improvement, established potential benefits, and communicated results to CCCSD personnel.

Task 3: Define Performance Measures

A tentative list of performance measures to support plant operations was developed by the performance measures task force. This list was distributed for comment to all department staff. Existing and new measures were evaluated and incorporated into new measures for plant operations.

Task 4: Establish Targets for Improvement

A memo was sent to all supervisors stating that individual work groups should develop tactics and metrics describing how they would meet plant operations performance measures.

Task 5: Implement Performance Measurement System

This task defined data and technical infrastructure requirements. These inputs will be incorporated into developing an Intranet site that has a dashboard look and feel.

Task 6: Measure and Initially Adjust

Plant operations performance measures will be deployed on the Intranet. Initial data will be gathered and evaluated. Changes will be incorporated into the plant operations intranet dashboard.

Task 7: Use the Performance Measurement System

Rollout of the intranet plant operations' dashboard will be completed. Reviews will be performed annually.

Education/Communication

- Monthly all-hands operations and maintenance staff meetings were held to highlight new management initiatives such as performance measures.
- Concept of performance measures and the balanced scorecard were introduced to CCCSD's performance measures task force.
- Draft performance measures were presented at an all-hands staff meeting.
- Additional training provided on balanced scorecard and performance measures to the performance measures task force and department managers and supervisors so they could better explain and support use.
- Memorandum was issued from CCCSD general manager to supervisor regarding strategic planning process and performance measure rollout. This memorandum stated that each workgroup should develop their own tactics and metrics for meeting performance measures.
- An article was written for the employee newsletter describing why focusing on CCCSD goals was needed and asked each employee for input on how to accomplish CCCSD's goals.

Lessons Learned

- Updating the strategic plan takes longer than expected.
- Employees need to see the relationship between performance measure activities and other utility activities. Enough time should be allowed in describing how new plan/goals relate to other programs.
- Performance measures need to be specific (i.e., not too general).
- Performance measures need to reflect and be balanced between the scorecard perspectives.

- Translation of perspectives into performance measures enhanced employees' understanding of what's important in day-to-day operations (e.g., minimal off-site odors).

- When linking performance measure to employee's personnel evaluations, it is important to be clear in how measures will be used to evaluate performance.

- While CCCSD management may set goals, achieving these goals and supporting goals requires buy-in and input at the workgroup level. This will also require roll-up metrics to a CCCSD-wide measurement.

- Much in the way of training, communication, and face-to-face meetings is necessary to get group participation in preparing metrics.

APPENDIX D

WEBSITES AND TECHNOLOGY EXAMPLES

D.1 Useful Websites

The following websites were consulted during the research and may be helpful to people implementing performance management systems:

www.2gc.co.uk—2GC Active Management

www.apqc.org/—American Productivity & Quality Center (APQC)

www.awwa.org/science/qualserve/—American Water Works Association – QualServe

www.schneiderman.com/Concepts/Scorecard/scorecard.htm—Arthur M. Schneiderman

www.rfabian.com/bsc/bsc.html—Balanced Scorecard article by Robert Fabian

www.bscol.com—The Balanced Scorecard Collaborative, Inc.

www.balancedscorecard.org—The Balanced Scorecard Institute

www.balancedscorecard.com—Balanced Scorecard Technology Council

www.quality.nist.gov—Baldrige National Quality Award

www.hbsp.harvard.edu/—Harvard Business Online

icma.org—International City/County Management Association (ICMA) Center for Performance Measurement

www.themanager.org/Knowledgebase/Management/BSC.htm—Management Balanced Scorecard

www.niwotridge.com/Resources/DomainLinks/BalancedScoreCard.htm—Niwot Ridge Resources

www.sas.com—SAS Strategic Performance Management / Balanced Scorecard

www.isixsigma.com—Six Sigma

www.targetimprovement.co.uk/—Target Business Improvement

www-mmd.eng.cam.ac.uk/—University of Cambridge, Institute for Manufacturing www.opm.gov/perform—Performance Management

www.zigonperf.com—Zigon Performance Group

APPENDIX E

EXAMPLE UTILITY SCORECARDS AND STRATEGY MAPS

E.1 Phoenix Water Services—Enterprise Scorecard

E.2 Phoenix Water Services—Tolerance Scorecard (Initiative-Based)

E.3 Union Sanitary District—Treatment and Disposal Scorecard (Process-Based)

E.4 Seattle Public Utilities
- Quarterly Metrics Report

E.5 Sample Strategy Maps
- Wastewater Collection (WEFTEC Workshop)
- Water Distribution (QualServe™ Workshop)
- Safety Strategy (Union Sanitary District)
- Operational Excellence Strategy (Union Sanitary District)
- T&D Process (Union Sanitary District)

E.1 Phoenix Water Services—Enterprise Scorecard

Customer – To achieve our vision, how should we appear to our customers?

Goal(s)	Objectives	Measures	Measure Target	Frequency of Measurement	Responsibility	Contact	FY 02/03 Data
High Quality, High Value Product	***Good taste and appearance with no odor***	Percent of customer days Taste & Odor measurements for Geosmine/MIB under the threshold	> 65%	Quarterly	Water Production / CRAO	Greenberg Hollander	65%
	Provide Safe Drinking Water	Percent of system days in 100% compliance with monitoring requirements and SDWA MCL's	100%	Semi-Annual	CRAO	Hollander	100%
Positive Memorable Experience	***Good neighbors - neighbors adjacent to or nearby a facility (e.g., visibility, aesthetics, weeds, smell, noise, etc)***	Number of Distribution, Collection, and Remote Facility complaints per year pertaining to site aesthetics	< 18	Quarterly	Collections / Distribution / Consumer Advocate Office	Ramon Puryear	12
	Courteous, respectful and presentable employees	Percent of negative responses from Distribution and Wastewater Collection Survey Cards	< 5.0%	Quarterly	Collections / Distribution	Ramon	3%
		Percent of positive responses from annual Customer Service Survey	> 70%	Annual	Customer Service	J. Smith	Office = 95% Field = 90%
Environmental Stewardship	***Resource reuse***	Combined total percent of effluent reused/recharged for all wastewater treatment plants	(> 75%) CC= > 85% 23rd= > 55% 91st= >75%	Quarterly	Water Production / Wastewater Treatment / CRAO	Walz (Lauer)	CCWRP = 75% 23rd Ave = 59% 91st Ave = 59%
	Pollution Prevention				Pollution Controls		
	Public education	Number of face-to-face contact hours with the public for education and program awareness purposes	?? Hrs	Quarterly	Water Conservation / Pollution Controls	Day Babcock Quay Kinshella	
	Regulatory compliance	Percent of plant days in 100% compliance with AZPDES discharge limits *	100%	Quarterly	CRAO	Hollander	99.60%
Consistent and Timely Service	***Readily accessible and available customer and product/service information***	Percent of received calls that are answered (Customer Service only)	> 95%	Quarterly	Customer Service	J. Smith	94%
	Seamless Service (flow of communication)						

* National AWWA Research Foundation Benchmark

Financial – To succeed financially, how should we appear to our "shareholders" (ratepayers, customers, public)?

Goal(s)	Objectives	Measures	Measure Target	Frequency of Measurement	Responsibility	Contact	FY 02/03 Data
Competitive Stable Rates	***Comparable rates (regional comparison)***	Within lower cost quartile of Southwest Group Rates (Using a combined Water/Wastewater monthly bill for Residential/Single Family/Average Usage)	Yes - lower cost quartile	Annual	Assistant Director's Office (Administration)	Blundon (D.Olson)	Within range
	Operate within budget	Percent of dollar difference between Department end-of-year actual costs versus appropriation	-1.0% to 0.0%	Annual	Management Services	Kellogg	-1.80%
Plan for Growth	***5 & 10 years CIP (including Water and Wastewater master plans, replacement and rehabilitation of existing infrastructure environmental regulations, growth)***	Percent of change order costs relative to total Capital Improvement Program value	< 7.0%	Semi-Annual	Planning and Programs	Goddard	10%
	Adequate financial reserves	Cash reserves as percentage of total annual expenditures (Water Fund)	> 12.5%	Annual	Assistant Director's Office (Administration)	Blundon (D. Olson)	
		Cash reserves as percentage of total annual expenditures (Wastewater Fund)	> 10.0%	Annual	Assistant Director's Office (Administration)	Blundon (D. Olson)	
Maintain Favorable Bond Rating	***Good ratings***	Moody's and S&P Bond Ratings	> /= AA	Annual	Assistant Director's Office (Administration)	Blundon (D. Olson)	Moody's = Aa S&P = AA
		Percent of Capital Improvement Program actual Bond Funds spent versus annual Capital Improvement Program cash flow	> 80%		Assistant Director's Office (Administration)	Blundon Kellogg	70%

Internal Business Process – To satisfy our customers, what business processes must we excel at?

Goal(s)	Objectives	Measures	Measure Target	Frequency of measurement	Responsibility	Contact	FY 02/03 Data
Exceptional Customer Service	***Bill accurately***	Number of billing adjustments made relative to 10,000 bills prepared *	< 65	Quarterly	Customer Service	J. Smith	146
Optimize Water and Wastewater Treatment, Collection and Distribution	***Cost efficient O&M***	Total direct and indirect costs per million gallons (MG) of drinking water processed (excludes Billing preparation costs)*	< $1,096 /MG	Quarterly	Water Production / Management Services	Kellogg Greenberg	$723/MG
		Total direct and indirect cost per million gallons (MG) of wastewater processed (excludes Billing preparation costs)*	< $1,140 /MG	Quarterly	Wastewater Treatment / Management Services	Kellogg Walz (Lauer)	$722/MG
	Minimal service disruption or outages	Percent of waterline leaks repaired within a 5 day service level	> 95%	Quarterly	Distribution	Ramon	93%
		Percent of water meter leaks repaired within a 3 day service level	> 95%	Quarterly	Customer Service	J. Smith	95%
Apply Available Resources Where Needed	***Budget appropriately***	Percent of difference between 0+12 estimate and end-of-year expenditure	-1.0% to 0.0%		Management Services	Kellogg	-0.7% under 0+12
Attract, Retain and Reward	***Focus on safety***	OSHA 300A - Health & Safety Severity Rate - Number of lost days per 200,000 hours *	< 100 Days	Quarterly	Human Resources	Davis	128 days

* National AWWA Research Foundation Benchmark

Learning & Growth – To achieve our vision, how will we sustain our ability to change and improve?

Goal(s)	Objectives	Measures	Measure Target	Method of Measurement	Frequency of Measurement	Responsibility	Contact	FY 02/03 Data
Developing, Maintaining, and Improving Skills Required to Provide State of the Art Services	***Comprehensive training to meet job requirements***	Average training hours per employee Number of training hours/Employee*	> 40 Hours per Employee		Quarterly	Director's Office	Nevitt	50.0 Hrs/Employee
	50 and 100 year assured water resources	Maintain designation of Assured Water Supply as issued by the Arizona Department of Water Resources	Yes					Present Designation = "Assured Supply"
High Performance Workplace	***Workforce Diversity***	Implementation of workforce diversity recommendations as measured by City Manager's Diversity Taskforce	> 8.0					Present Rating = 7.3
Explore & Implement New Ways to do Business	***Explore alternate service delivery methods***	Percent of Capital Improvement Project funds used/dedicated for alternate delivery methods (DBO/CM@Risk)	??%		Annual	Planning and Programs	Goddard	
		Job Order Contracting	>65%					
	Encourage research and innovation (external focus)	Qual-Serve Organizational Best-Practices Index (Self-Scored) *	> 25		Annual	Assistant Director's Office (Operations, Technical Services, Administration, Development and Planning, Director's Office)	Blundon	26

* National AWWA Research Foundation Benchmark

E.2 Phoenix Water Services—Zero Tolerance Scorecard (Initiative-Based)

Customer – To achieve our vision, how should we appear to our customers?											
Goal(s)	**Objectives**	**Business Process**	**Measures**	**Measure Target**	**Method of Measurement**	**Frequency of Measurement**	**Responsibility**	**Applicability** (D=Direct; I=Indirect; X=N/A)	**Improvement Prioritization**		
									Qual	**Effec**	**Effic.**
Positive Memorable Experience	*Good neighbors*	Overflow Incident processes	Spill response time	< 2 hours	Spill Response Report	1 per month	Wastewater Collections/ CRAO	D		✓	
Environmental Stewardship	*Pollution prevention*	Manhole Entry Permitting Program	# of requests reviewed	90% reviewed within 10 working days	Weekly Report	Weekly	PCD	D		✓	
		New Water Account Pretreatment Survey	# of new water account surveys performed	90% reviewed within 90 days	Weekly Report	Weekly	PCD				
		New Interceptor Plans and Specifications Review	1200 per year	80% completed within 30 days	Weekly Report	Weekly	PCD	D		✓	
		Interceptor Construction Inspection	400 per year	90% inspected within 24 hours of request	Weekly Report	Weekly	PCD	D		✓	
	Regulatory compliance	Reporting	% of on time required reports	100% ontime	Weekly Report	Weekly	CRAO	D	✓		
		Enforcement Initiation & Response	Return to Compliance (Inputs: # of NOVs issued/ # of complaint resolutions)	100% compliance within 7 calendar days of NOV issuance or *within specified deadline* to allow for construction	Weekly Report	Weekly	PCD				
Consistent and Timely Service	*Readily Accessible and available customer and product/service information*	Public Protection	# of hours to initiate cleanup	< 1 hour after blockage cleared	Spill Response Report	Monthly	Wastewater Collections/ CRAO	D		✓	✓
			# of hours to cleanup	< 3 hour after blockage cleared	Spill Response Report	Monthly	Wastewater Collections/ CRAO	D		✓	✓

Financial – To succeed financially, how should we appear to our "shareholders" (ratepayers, customers, public)?											
Goal(s)	**Objectives**	**Business Process**	**Measures**	**Measure Target**	**Method of Measurement**	**Frequency of Measurement**	**Responsibility**	**Applicability** (D=Direct; I= Indirect; X=N/A)	**Improvement Prioritization**		
									Qual	**Effec**	**Effic**
Plan for Growth	*5, 10, & 20 years CIP (including W/WW master plans, replacement and rehabilitation of existing infrastructure environmental regulations, growth)*	Construction Inspection/ Oversight	Funding in CIP	Approval by E&R (Budget & Research)/ Council	Official CIP	2/Year	MSD/ Wastewater Engineering	D	✓	✓	
		Design Evaluations & Hydraulic Modeling	Funding in CIP	Approval by B&R (Budget & Research)/ Council	Official CIP	2/Year	MSD/ Wastewater Engineering				

Internal Business Process – To satisfy our customers, what business processes must we excel at?											
Goal(s)	**Objectives**	**Business Process**	**Measures**	**Measure Target**	**Method of Measurement**	**Frequency of measurement**	**Responsibility**	**Applicability** (D=Direct; I=Indirect; X=N/A)	**Improvement Prioritization**		
									Qual	**Effec**	**Effic**
Exceptional Customer Service	*Provide customer desired and needed services*	Educate Customers/ Contractors/ Consultants/ Vendors	Customer Education Survey	350 per year	Weekly Report	Weekly	PCD	D	✓	✓	
		Outreach Events	# of events	12 events per year	Weekly Report - Date of Event	Weekly	PCD				
		Retain/ Containment of Flow Where Possible	Overflows that reach waters of the U.S.	0 overflows that reach waters of the U.S.	Spill Response Report	Quarterly	CRAO				
		Clear Blockages	Time to clear blockages	100% cleared within 8 hours	Spill Response Report	Monthly	Wastewater Collections				
Optimize W/WW Treatment, Collection/ Distribution	*Effective and cost efficient, O&M*	Sewer Cleaning	# of miles cleaned (15 inches or under)	100% of system every 3 [or more?] years	City Managers Monthly Report	Monthly	Wastewater Collections	D	✓		✓
		Sewer Inspections	# of miles inspected (<15 inches)	100% of system every 3 [or more?] years	Weekly Summary Productivity Report	Monthly	Wastewater Collections				
		Manhole inspections	# manholes inspected	100% of system every 3 [or more?] years	Weekly Summary Productivity Report	Monthly	Wastewater Collections				
		Equipment Monitoring, PM	# of pumps inspected/rebuilt	100 % every 6 months/as needed or every 5 years	Job Reports	Monthly	Wastewater Collections				
		Repair Piping or Manholes (WWC)	# repaired before SSO	100% repaired	Job Reports; Weekly Summary Productivity Report	Monthly	Wastewater Collections				
		Rehabilitate Lines or Manholes (WWC)	# rehabilitated before SSO	100% rehabilitated	Job Reports; Weekly Summary Productivity Report	Monthly	Wastewater Collections				
		Sewer Replacement/ Rehabilitation (WWE)	Follow priority ranking of sewer mains for replacement/rehabilitation	0 SSOs from main collapses	CIP/Sewer Survey Reports	Monthly	Wastewater Engineering				

Internal Business Process – To satisfy our customers, what business processes must we excel at?											
Goal(s)	**Objectives**	**Business Process**	**Measures**	**Measure Target**	**Method of Measurement**	**Frequency of measurement**	**Responsibility**	**Applicability** (D=Direct; I=Indirect; X=N/A)	**Improvement Prioritization**		
									Qual	**Effec**	**Effic**
Apply Available Resources Where Needed	*Inter-department, inter-division effective communication and allocation*	Coordination of Emergency Response Plan Implementation	# of Spill Response forms' Part A submitted by next business day	100% completion	Spill Response Report (E mail date and time)	Weekly	Wastewater Collections/ CRAO	D		✓	✓
			Verbal notification to CRAO within 4 hours of trigger event	100%	Voice mail time and Spill Response Report	Weekly	Wastewater Collections/ CRAO				
		Incidence Response Performance	Incidence Response Performance Debriefing meeting	Once per quarter for selected events	Weekly Report	Weekly	CRAO, Wastewater Collections, PCD				
Attract, Retain and Reward	*Focus on safety*	Training	# of employees completing yearly training	100%	Attendance Rosters/ CHRIS	Annual Based on Calendar Year	Training	D		✓	

Learning & Growth – To achieve our vision, how will we sustain our ability to change and improve?

Goal(s)	Objectives	Business Process	Measures	Measure Target	Method of Measurement	Frequency of Measurement	Responsibility	Applicability (D=Direct; I=Indirect; X=N/A)	Improvement Prioritization		
									Qual	Effec	Effic
Explore & Implement New Ways to do Business	*Encourage research and innovation (external focus)*	Research &/or Apply Best New Practices	Sewer Corrosion Prevention (MAG) - maintain target pH	pH > 6	pH Test Results	every 6 months or 1 year depending on pH maintenance > 6 (increase frequency if pH less than 6)	Wastewater Engineering	D		✓	
		Spill Response Investigation and prevention	# investigated and mapped [*future* - causes identified]	100%	Spill Response Report	Monthly	PCD/ CRAO				
	Implement environmentally sustainable business practices (internal focus)	Research &/or Apply Best New Practices	Policy to engage in research projects and review or employ best practices - # of meetings per year	1 meeting per year	Weekly Report	Monthly	CRAO	D		✓	

E.3 Union Sanitary District—Treatment and Disposal Scorecard (Process-Based)

Objectives	Measures
Customer Perspective: Operate & maintain the Plant 24/365 (uninterrupted service)	Number of priority 1 (emergency) work orders Number of adverse impacts (odor complaints, violations)
Financial Perspective: Optimize T&D operating costs: Minimize energy and chemical use Improve utilization of labor	Total KWh/MG, Electricity cost, value of electricity generated on site Cost of hypochlorite and polymer % total work hours spent on maintenance work Overtime as % of base payroll
Internal Process Perspective: Maintain Plant processes within parameters	Plant operational health index • Monthly index value • # days <75%
Implement an effective predictive/preventative maintenance program	% maintenance time spent on planned vs. unplanned activities % of priority 1 work orders completed within 3 days % of priority 2 work orders completed within 30 days
Learning and Growth: Develop program for multi-skilled workers Enhance employee skills	# of job competency standards developed % employees demonstrating competence in each area

Safety Strategy (Union Sanitary District)

Objectives	Measures	Initiatives
Maintain compliance	# Potential violations identified during most recent CSRMA audit #addressed/resolved	Implement corrective actions
Maintain a staff that is present and healthy and able to work, in order to meet customer needs	Ave full-time equivalent (FTE) lost time Ave FTE limited duty time	Coordination of safety training FY04 FY05 Training Plan Safety Committee SOP development Accident investigations
	# New OSHA recordable accidents Accidents involving lost or limited-duty time	
Reduce costs of accident/injuries	Insurance modifier rate (x-mod) Total cost of lost time Total cost of limited duty time	
Encourage safe behaviors	# Management observation visits with feedback per quarter	Field observation program
Provide a safe work place	# Facility inspections performed per quarter	Facility inspection program
Increase employee skills (management and staff)	# Safety training programs offered % affected employees trained	FY 05 Training Plan
Promote a positive safety culture	Results of periodic employee survey	Survey June 2005 (follow-up to 2004 survey)

E.4 Seattle Public Utilities—Quarterly Metrics Report Water LOB

Section #	Metric #/ Name	Definition	Target	Reporting Frequency
1	1: O&M Projected Net Income	Current estimate of year-end Net Income (Total Revenue – Total Costs)	\$11.9 million	Quarterly
1	2: Operating Cash Position	Water Fund's projected operating cash balance at year-end	\$4.4 million	Quarterly
2	3: Retail Bill Comparison	A rank comparison of a residential monthly bill at 6ccf of usage between Seattle and eight peer jurisdictions (San Francisco, East Bay MUD, San Diego, Portland, Bellevue, Tacoma, Everett, Columbus)	5	Annual
2	4: Percent of Revenue Used for Debt Service	Percentage of total operating revenue used for the payment of principal and interest on debt	TBD	Annual
2	5: Bond Rating	Average of current Moody's and Standard and Poor's rating on 1st lien debt	AA	Annual
3	6: Primary Health Standards	Compliance with primary health standards administered by the WDOH Drinking Water Program	100%	Annual
3	7: Water Pressure	Number of customers experiencing a pressure incident (below 30 psi) at the point of connection of the property to the utility's main during the peak hour other than as a result of a water interruption, during a fire, or temporary operational problems.	2,700 properties or less (1.5%)	Annual
3	8: Water Outage	Number of customers experiencing a cumulative outage of water from one or more events totaling greater than 4 hours in a year	7,200 customers or less annually (4%)	Annual
4	9: New Water Service Installations	Percentage of taps not requiring permits completed within 28 days of customer payment, unless a delay is requested by the customer	95%	Annual
5	10: Conservation	Total annual permanent and temporary MGD reduction. Goal is to save 18 MGD by 2010 by reducing consumption by 1% per year	TBD	Annual

Section #	Metric #/ Name	Definition	Target	Reporting Frequency
5	11: Real Losses	Percentage of water losses due to SPU system leakages (i.e., actual water lost through breaks/leaks in the distribution system)	5% or less	Annual
5	12: Long-Range Supply Availability	Number of years until forecasted average annual water demand equals firm yield from current and planned supply sources	20 years or more	Reviewed annually
5	13: 6-Month Water Supply Outlook	Percentage of normal (average) water supply availability based on (depending on the season) reservoir supply, mountain snowpack, and inflows into the reservoirs	70%	Updated quarterly in mid-January, April, July, and October
6	14: Instream Flows	A composite of 8 in-stream flow protection measures totaling no more than 100 points. Excellent = 90 to 100 pts; Good = 80 to 90 pts; Marginally Acceptable = 70 to 80 pts; Needs Substantial Improvement = 60 to 70 pts; Unacceptable = less than 60 pts		

E.5 Sample Strategy Maps

Wastewater Collection (WEFTEC Workshop)

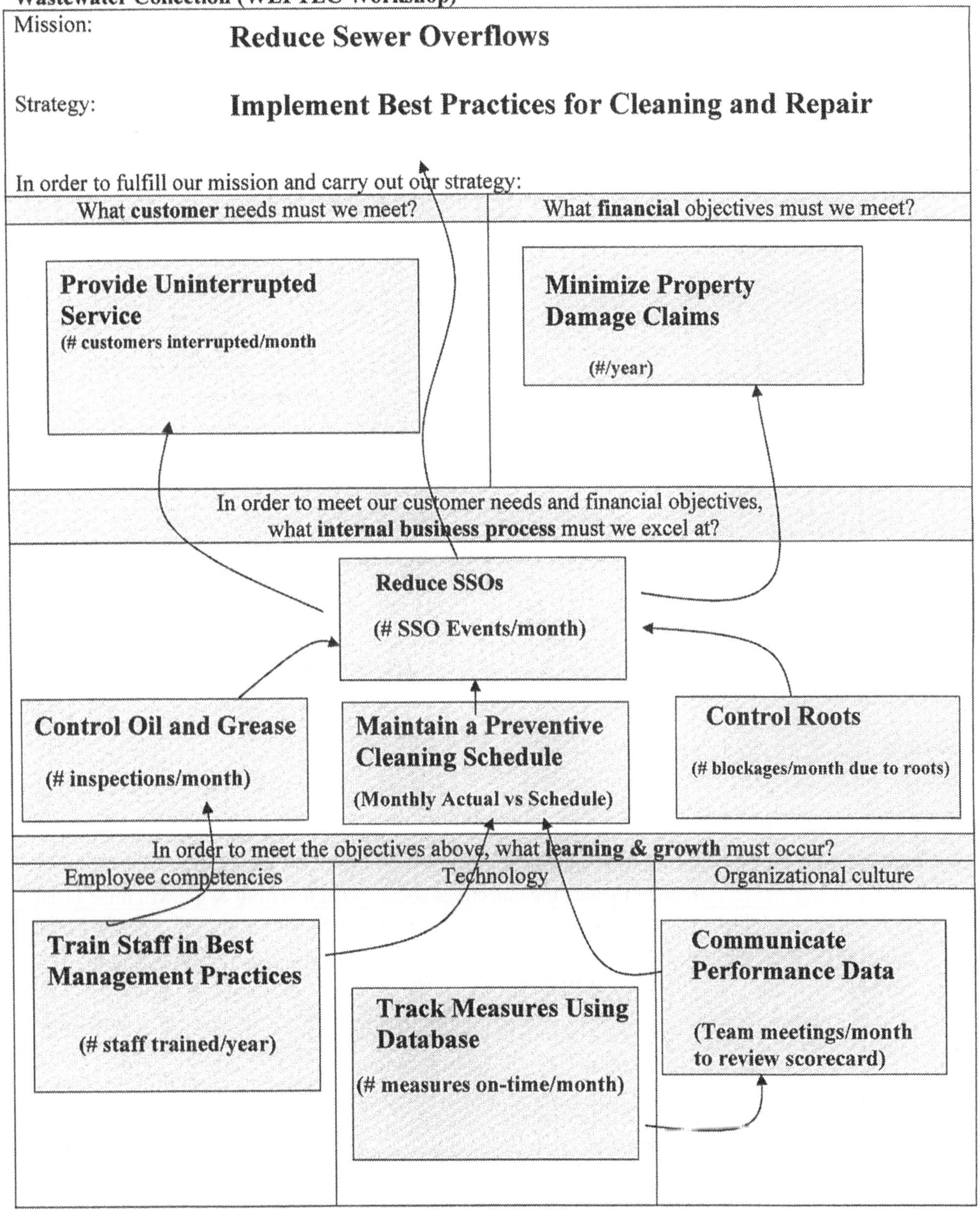

Wastewater Distribution (QualServe Workshop)

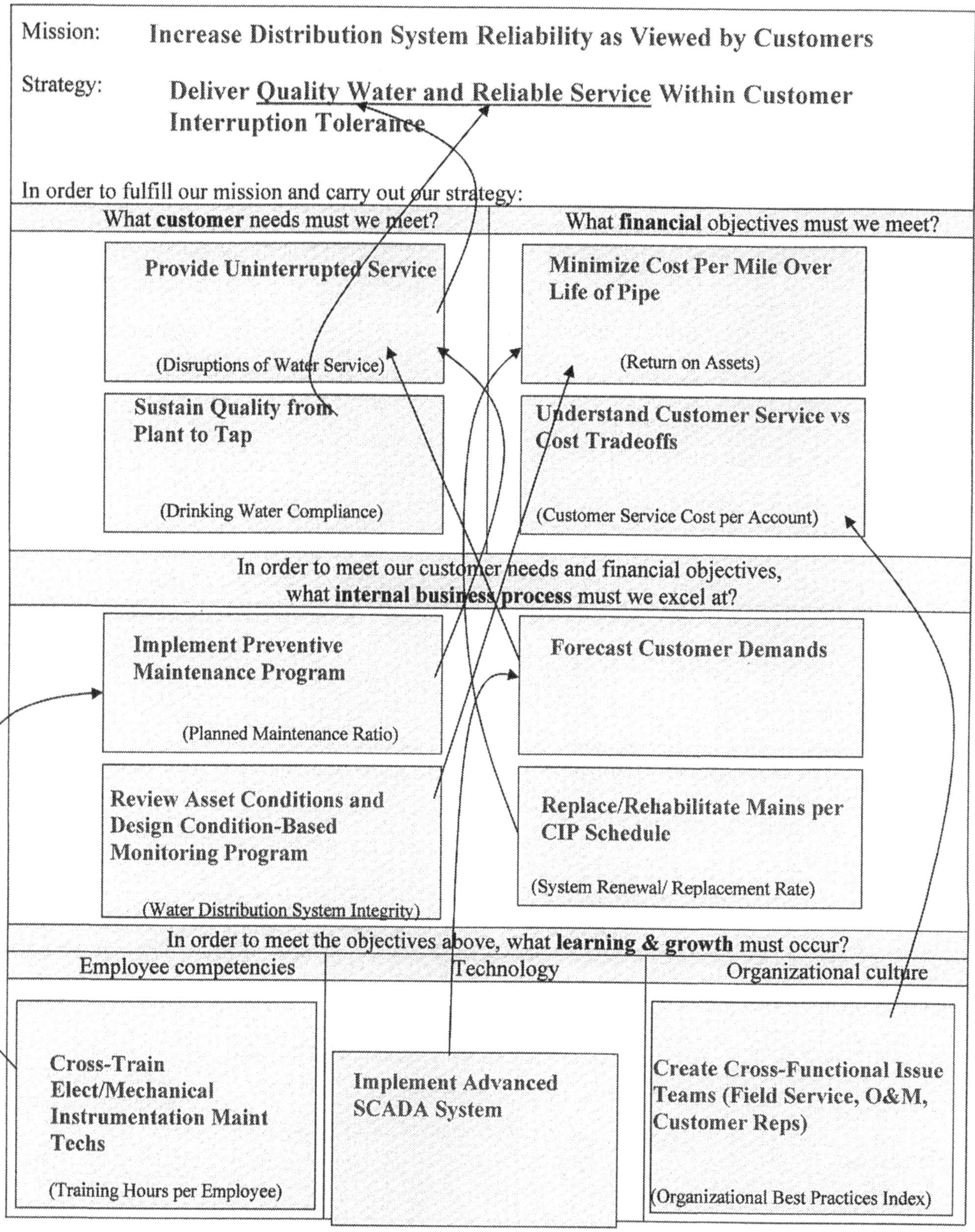

Safety Strategy (Union Sanitary District)

The Safety Scorecard uses elements of the Safety Incentive Pilot Program to focus attention on improving the injury and accident rates at the District. All teams include their safety incentive baseline score and their share of the cost of accidents on their team scorecard (regardless of whether they've developed an action plan). Business Services calculates the cost of injuries and accidents at the District and distributes the information to each team.

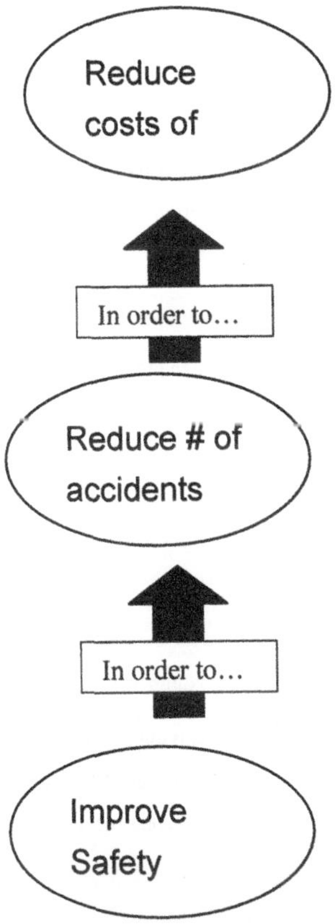

Objectives	Measurements
Financial: ♦ Reduce costs of accidents	♦ Total costs ♦ % Improvement
Internal processes: ♦ Use TQ tools to reduce the number of accidents	♦ Team safety incentive scores ♦ % Improvement
Learning and Growth: ♦ Improve Safety Awareness	♦ Number of teams with safety action Plans

Operational Excellence Strategy (Union Sanitary District)

The Operational Excellence strategy focuses on training, process improvements, financial performance, and customer responsiveness needed to accomplish USD's Mission. Objectives were taken from the original District scorecard, only narrowed to better reflect the District priorities over the next few years.

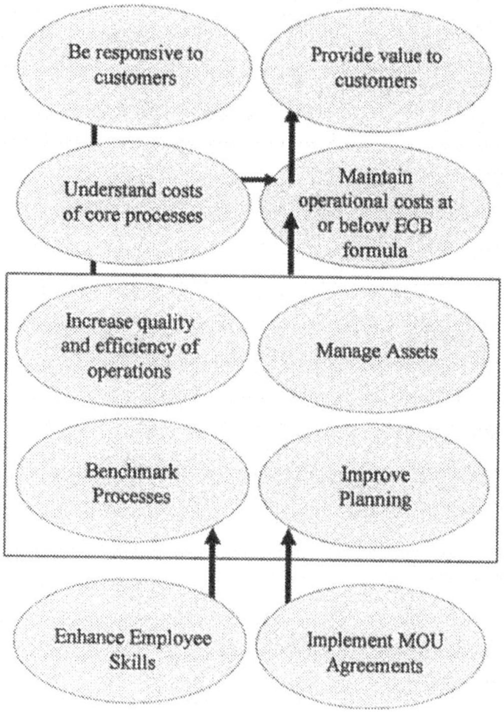

Objectives	Measurements
Customer:	
• Be responsive to customers (problems, complaints)	• Number of adverse impacts our activities have on external customers
• Provide value to customers	• USD rates as compared to surrounding area
Financial:	
• Maintain operational costs at or below ECB formula	• ECB Expenditures Actual % of budget
• Understand costs of core processes	• Progress developing a cost allocation method by core processes
Internal Processes:	
• Increase efficiency of operations • Benchmark: compare processes and costs with others and ourselves over time • Manage District assets • Improve planning	• See process and support scorecards
Learning and Growth:	
• Enhance employee skills • Implement MOU agreements affecting teams	• Number of employees trained in business/budgeting and problem-solving skills • Transfer of training rates • Progress meeting contact milestones

T&D Process (Union Sanitary District)

Process Scorecard changes the way T&D measures performance: a cross-functional team of Treatment and Disposal employees built a T&D process scorecard. The scorecard was developed in order to focus attention at the process level and integrate several previous improvement efforts, including the AMSA competitiveness assessment. The scorecard focuses on opportunities for improvement, including automation of plant processes, increasing employees' skills, and reducing chemical and energy costs.

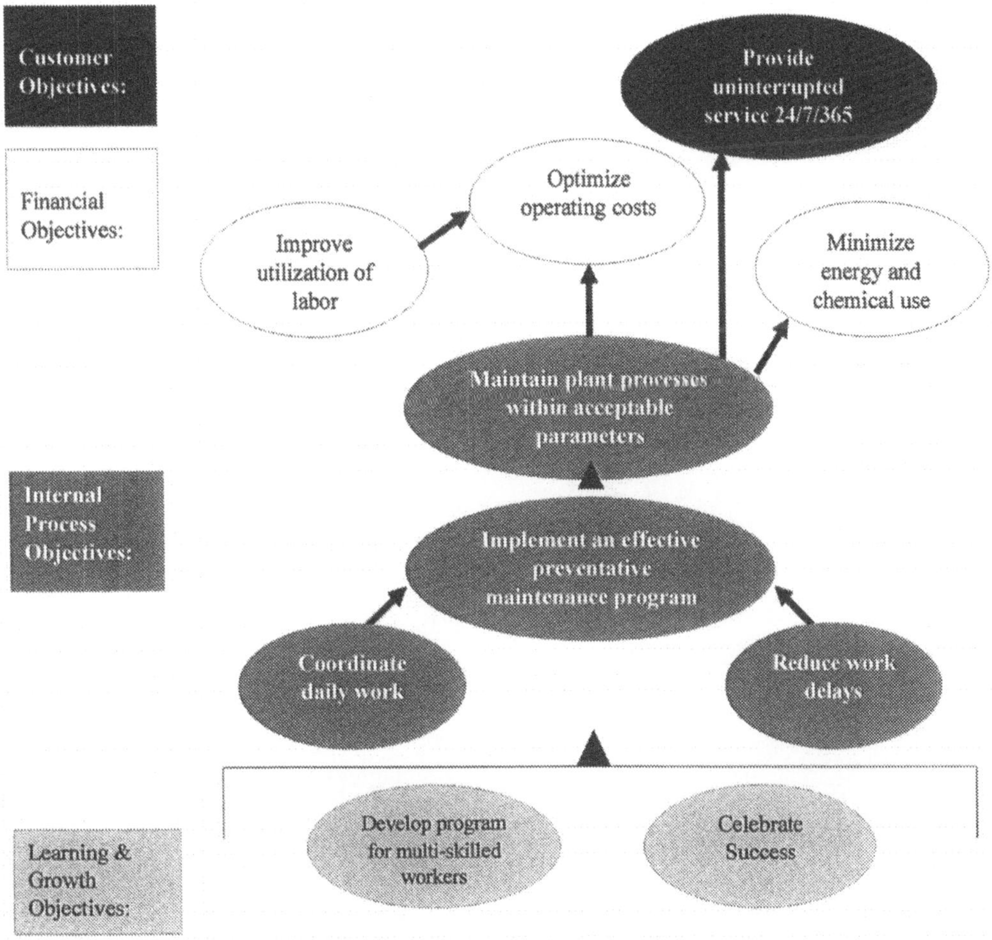

APPENDIX F

UTILITY BUSINESS PROCESS FRAMEWORK

F.1 Framework Development

Framework Categories

To define a utility's overall services, categories of processes have been split into two types: core services and business services. The cores services are chains of processes extending from customer needs to the environment that provide additional value with each subsequent process, hence forming value chains. This definition of service is based on original work under a project for AwwaRF (1997, the utility business architecture [UBA]). The business processes do not fit directly into this chain of processes. Instead they support the value chain processes, usually more than one. For example, provision of vehicles and equipment—a business process — supports many other business processes and core processes, such as production and delivery of water, collection and treatment of wastewater, customer service, and laboratory. The definition of business processes is based on an adaptation of a process classification framework (APQC, 1992) to fit typical wastewater and water utilities.

Organizational Independent Framework

Since one or more processes may be assigned to a business unit, the business process framework (UBA) is organizationally independent. It is also possible for a process to be carried out by more than one business unit.

Subprocesses have been identified for each process. Some subprocesses are repeated in multiple places in order to define the processes completely and independently.

Three subprocesses in particular are often repeated in the process definition but may be performed in various places in the organizational structure. The subprocesses are:

Acquire Resources

The subprocess to plan for and acquire necessary resources appears in several processes. This subprocess is often split between two or more business units. One unit is responsible for planning the necessary resources, another for coordinating budgets, while yet another business unit (e.g., procurement, human resources, or engineering) actually obtains the resources. Specialized processes of fleet, facilities, and information technology respectively provide the resources of transportation, facilities (buildings and grounds), and computers. The purpose of listing this subprocess in several of the framework processes is to show that the responsibility to plan for and acquire necessary resources is an integral part of the process.

Maintenance

The subprocess to schedule and perform maintenance appears in several of the processes, either alone or as a partner with manage processes (operate physical assets). The intent of multiple listings is to show that the process is responsible for maintenance as well as operation of its physical assets. An organizational structure is not implied. In some organizations, the operations and routine maintenance for one or more processes could be assigned to the same business unit. In other instances, a separate business unit or external resource could provide part or all of the maintenance.

Ensure Product Quality

The third subprocess to appear in multiple definitions is ensure product quality. The need to ensure product quality is an integral part of the product-oriented processes. While a separate organizational unit such as a laboratory may perform portions of this subprocess, the responsibility for quality remains with the processes. Tasks of this subprocess include:

- Identify testing needs and requirements
- Track samples
- Perform tests
- Report test results
- Interpret test results

Service and Process Classification

The typical water and wastewater core and business services and the related subprocesses are listed on the following service and process decomposition.

UBA Core Services

Produce and Deliver Water

Source Protection (Protect Water Sources)

- Plan for and acquire necessary resources
- Protect sources from degradation
- Manage processes
- Schedule and perform maintenance
- Ensure product quality

Water Production (Produce Water)

- Plan for and acquire necessary resources
- Abstract necessary raw water quantity
- Manage processes
- Schedule and perform maintenance
- Ensure product quality

Transmission (Transmit Water)

- Plan for and acquire necessary resources
- Deliver necessary raw water quantity
- Manage processes
- Schedule and perform maintenance
- Ensure product quality

Treatment (Treat Water)

- Plan for and acquire necessary resources
- Treat water to required quality
- Manage processes
- Schedule and perform maintenance
- Ensure product quality

Distribution (Distribute Water)

- Plan for and acquire necessary resources
- Deliver necessary water quantity
- Manage processes
- Schedule and perform maintenance
- Ensure product quality

Collect and Treat Wastewater

Environment Monitoring (Monitor Environment)

- Plan for and acquire necessary resources
- Measure environment for background levels
- Manage processes
- Schedule and perform maintenance
- Ensure product quality

Solids Disposition (Dispose of Solids)

- Plan for and acquire necessary resources
- Treat wastewater solid streams
- Dispose or apply biosolids
- Manage processes
- Schedule and perform maintenance
- Ensure product quality

Treatment (Treat Wastewater)

- Plan for and acquire necessary resources
- Treat wastewater liquid streams
- Manage processes

- Schedule and perform maintenance
- Ensure product quality

Wastewater Collection (Collect Wastewater)

- Plan for and acquire necessary resources
- Collect wastewater (gravity and pumped)
- Manage processes
- Schedule and perform maintenance
- Ensure product quality

Sewer Use Monitoring (Monitor Sewer Use)

- Plan for and acquire necessary resources
- Measure discharges for overstrength/wastes and violations
- Manage processes
- Schedule and perform maintenance
- Ensure product quality

UBA Business Processes

Processes to Acquire Resources

Engineering (Plan, Design, and Construct Facilities and Infrastructure)

- Identify capital improvement priorities
- Develop a plan to repair and/or replace infrastructure
- Design facilities/infrastructure
- Construct facilities/infrastructure
- Evaluate systems and monitor systems performance

Facilities (Manage Real Estate and Facilities)

- Develop a land and facilities use strategy plan
- Acquire or lease facilities
- Maximize use of facilities
- Acquire and dispose needed and surplus lands
- Manage buildings and grounds

Fleet (Provide Vehicles and Equipment)

- Manage transportation needs
- Distribute vehicles and nonstationary equipment
- Process, operate and maintain fleet vehicles and equipment

Human Resources (Manage and Develop Human Resources)

- Create and manage human resource strategies
- Manage deployment of personnel

- Develop and train employees
- Manage employee performance, reward, and recognition
- Ensure employee well-being and satisfaction
- Manage labor-management relations

Information Technology (Manage Information Systems)

- Plan for information resources management
- Develop an IT strategic plan
- Develop and deploy information management systems
- Manage information storage and retrieval
- Manage facilities and network operations
- Provide information systems support services
- Facilitate information sharing and communications
- Evaluate and audit information quality

Procurement (Procure, Contract, and Inventory Goods & Services)

- Maintain vendor lists
- Plan and manage the procurement process
- Establish preferred suppliers and blanket orders
- Prepare and issue RFPs and RFQs
- Monitor performance of contract recipients
- Manage stores/distribution transactions (inventory control)

Other Processes

Customer Service (Service Customers)

- Meter reading and consumption/discharge quantification
- Bill for services and consumption/discharge
- Collection—process payment remittance and delinquent accounts
- Provide service initiation/discontinuation
- Provide customer field services to complete service orders
- Respond to customer inquiries and complaints

Planning (Develop Business Plans and Strategy)

- Monitor the external market environment
- Establish vision and mission
- Define the business plan and organizational strategy
- Design the organizational structure and relationship between organizational units
- Develop and set goals and targets
- Develop long-range capacity and resource plans
- Develop the capital improvement plan (capital planning and response)

External Relations (Maintain External and Stakeholder Relations)

- Communicate regularly with the media and/or stakeholders
- Manage government and agency relationships
- Build funding relationships
- Develop targeted public relations programs
- Manage legal and ethical issues
- Manage other utility relationships

Marketing (Understand Markets and Customers, Set Rates, Sell Service/Conservation)

- Determine customer needs
- Measure customer satisfaction
- Monitor changes in market and/or customer satisfaction
- Market services by customer class
- Sell services to nontraditional utility customers
- Create and sell innovative products/services
- Forecast costs of service and ratemaking
- Establish rates and fees

Finance (Manage Finance and Accounting)

- Plan and manage financial resources
- Process financial and accounting transactions
- Report financial information
- Conduct internal audits
- Manage the tax function
- Produce annual budgets
- Produce annual financial reports
- Develop bond offerings for debt finance

Compliance and Risk (Manage Regulatory Compliance and Emergency Response)

- Develop an environmental management strategy
- Monitor environmental performance
- Ensure compliance with regulations
- Manage water conservation
- Manage remediation efforts
- Implement emergency response programs (emergency planning response)
- Manage health and safety programs

Legal (Provide Legal Services and Risk Management)

- Mix internal and external counsel
- Monitor and controlling external legal costs
- Conduct risk and vulnerability assessments
- Develop a risk management strategy

- Identify, assess, and address risk versus loss
- Evaluate viable, cost-effective alternatives
- Analyze premiums paid

Performance Improvement (Measure Performance and Continuously Improve)

- Measure organizational performance
- Conduct quality assessments
- Benchmark performance
- Improve processes and systems
- Implement improvement plans

WASTEWATER UTILITY

Alabama
Montgomery Water Works & Sanitary Sewer Board

Alaska
Anchorage Water & Wastewater Utility

Arizona
Gila Resources
Glendale, City of, Utilities Department
Mesa, City of
Peoria, City of
Pima County Wastewater Management

Arkansas
Little Rock Wastewater Utility

California
Calaveras County Water District
Central Contra Costa Sanitary District
Crestline Sanitation District
Delta Diablo Sanitation District
Dublin San Ramon Services District
East Bay Dischargers Authority
East Bay Municipal Utility District
Eastern Municipal Water District
El Dorado Irrigation District
Fairfield-Suisun Sewer District
Fort Lauderdale, City of
Irvine Ranch Water District
Las Virgenes Municipal Water District
Livermore, City of
Lodi, City of
Los Angeles, City of
Los Angeles County, Sanitation Districts of
Napa Sanitation District
Orange County Sanitation District
Palo Alto, City of
Riverside, City of
Sacramento Regional County Sanitation District
San Diego Metropolitan Wastewater Department, City of
San Francisco, City & County of
San Jose, City of
Santa Barbara, City of
Santa Cruz, City of
Santa Rosa, City of
South Bayside System Authority
South Coast Water District
South Orange County Wastewater Authority
Sunnyvale, City of
Union Sanitary District
West Valley Sanitation District

Colorado
Boulder, City of
Colorado Springs Utilities
Greeley, City of
Littleton/Englewood Water Pollution Control Plant
Metro Wastewater Reclamation District, Denver

Connecticut
The Mattabassett District
New Haven, City of, WPCA

District of Columbia
District of Columbia Water & Sewer Authority

Florida
Broward, County of
Fort Lauderdale, City of
Gainesville Regional Utilities
JEA
Miami-Dade Water & Sewer Authority
Orange County Utilities Department
Reedy Creek Improvement District
Seminole County Environmental Services
St. Petersburg, City of
Stuart Public Utilities
Tallahassee, City of
Tampa, City of
Toho Water Authority
West Palm Beach, City of

Georgia
Atlanta Department of Watershed Management
Augusta, City of
Clayton County Water Authority
Cobb County Water System
Columbus Water Works
Fulton County
Gwinnett County Department of Public Utilities
Savannah, City of

Hawaii
Honolulu, City & County of

Idaho
Boise, City of

Illinois
American Bottoms Wastewater Treatment Plant
Greater Peoria Sanitary District
Kankakee River Metropolitan Agency
Metropolitan Water Reclamation District of Greater Chicago
Wheaton Sanitary District

Iowa
Ames, City of
Des Moines Metro Wastewater Reclamation Authority
Iowa City

Kansas
Johnson County Unified Wastewater Districts
Unified Government of Wyandotte County/ Kansas City, City of

Kentucky
Louisville & Jefferson County Metropolitan Sewer District

Louisiana
Sewerage & Water Board of New Orleans

Maine
Bangor, City of
Portland Water District

Maryland
Anne Arundel County Bureau of Utility Operations
Howard County Department of Public Works
Washington Suburban Sanitary Commission

Massachusetts
Boston Water & Sewer Commission
Upper Blackstone Water Pollution Abatement District

Michigan
Ann Arbor, City of
Detroit, City of
Holland Board of Public Works
Lansing, City of
Saginaw, City of
Wayne County Department of Environment
Wyoming, City of

Minnesota
Rochester, City of
Western Lake Superior Sanitary District

Missouri
Independence, City of
Kansas City Missouri Water Services Department
Little Blue Valley Sewer District
Metropolitan St. Louis Sewer District

Nebraska
Lincoln Wastewater System

Nevada
Henderson, City of

New Jersey
Bergen County Utilities Authority
Passaic Valley Sewerage Commissioners

New York
New York City Department of Environmental Protection

North Carolina
Charlotte/Mecklenburg Utilities
Durham, City of
Metropolitan Sewerage District of Buncombe County
Orange Water & Sewer Authority

Ohio
Akron, City of
Butler County Department of Environmental Services
Columbus, City of
Metropolitan Sewer District of Greater Cincinnati
Northeast Ohio Regional Sewer District
Summit, County of

Oklahoma
Trinity River Authority
Tulsa, City of

Oregon
Clean Water Services
Eugene/Springfield Water Pollution Control
Water Environment Services

Pennsylvania
Philadelphia, City of
University Area Joint Authority

South Carolina
Charleston Commissioners of Public Works
Mount Pleasant Waterworks & Sewer Commission
Spartanburg Sanitary Sewer District

Tennessee
Cleveland, City of
Knoxville Utilities Board
Murfreesboro Water & Sewer Department
Nashville Metro Water Services

Texas
Austin, City of
Dallas Water Utilities
Denton, City of
El Paso Water Utilities
Fort Worth, City of
Gulf Coast Waste Disposal Authority
Houston, City of
San Antonio Water System
Trinity River Authority

Utah
Salt Lake City Corporation

Virginia
Alexandria Sanitation Authority
Arlington, County of
Fairfax County
Hampton Roads Sanitation District
Henrico, County of

WERF SUBSCRIBERS

Hopewell Regional Wastewater Treatment Facility
Loudoun County Sanitation Authority
Lynchburg Regional WWTP
Prince William County Service Authority
Richmond, City of
Rivanna Water & Sewer Authority

Washington
Bellevue Utilities Department
Edmonds, City of
Everett, City of
King County Department of Natural Resources
Seattle Public Utilities
Sunnyside, Port of
Yakima, City of

Wisconsin
Green Bay Metro Sewerage District
Madison Metropolitan Sewerage District
Milwaukee Metropolitan Sewerage District
Racine, City of
Sheboygan Regional Wastewater Treatment
Wausau Water Works

Australia
South Australian Water Corporation
Sydney Water Corporation
Water Corporation of Western Australia

Canada
Greater Vancouver Regional District
Regina, City of, Saskatchewan
Toronto, City of, Ontario
Winnipeg, City of, Manitoba

Mexico
Servicios de Agua y Drenaje de Monterrey, I.P.D.

New Zealand
Watercare Services Limited

United Kingdom
Yorkshire Water Services Limited

STORMWATER UTILITY

California
Los Angeles, City of, Department of Public Works
Monterey, City of
San Francisco, City & County of
Santa Rosa, City of

Colorado
Boulder, City of

Georgia
Griffin, City of

Iowa
Cedar Rapids Wastewater Facility
Des Moines Metro Wastewater Reclamation Authority

Kansas
Overland Park, City of

Maine
Portland Water District

North Carolina
Charlotte, City of, Stormwater Services

Pennsylvania
Philadelphia, City of

Tennessee
Chattanooga Stormwater Management

Washington
Seattle Public Utilities

STATE

Arkansas Department of Environmental Quality
Fresno Department of Public Utilities
Kansas Department of Health & Environment
Urban Drainage & Flood Control District, CO

CORPORATE

ADS Environmental Services
The ADVENT Group Inc.
Alan Plummer & Associates
Alden Research Laboratory Inc.
Aqua-Aerobic Systems Inc.
Aquateam–Norwegian Water Technology Centre A/S
ARCADIS
Associated Engineering
BaySaver Inc.
Black & Veatch
Blasland, Bouck & Lee Inc.
Boyle Engineering Corporation
Brown & Caldwell
Burns & McDonnell
CABE Associates Inc.
The Cadmus Group
Camp Dresser & McKee Inc.
Carollo Engineers Inc.
Carpenter Environmental Associates Inc.
CDS Technologies Inc.
Chemtrac Systems Inc.
CH2M HILL
Damon S. Williams Associates, LLC
David L. Sheridan, P.C.
Dewling Associates Inc.
Earth Tech Inc.
Ecovation
EMA Inc.
The Eshelman Company Inc.
EWT Holdings Corporation
Fay, Spofford, & Thorndike Inc.
Freese & Nichols Inc.
ftn Associates Inc.
Fuss & O'Neill Inc.
Gannett Fleming Inc.
GE Betz
Greeley and Hansen LLC
Hazen & Sawyer, P.C.
HDR Engineering Inc.
HNTB Corporation
HydroQual Inc.
Infilco Degremont Inc.
Ingersoll-Rand Energy Systems
Insituform Technologies Inc.
Institute for Environmental Technology & Industry, Korea
Jacobson Helgoth Consultants Inc.
Jason Consultants LLC Inc.
Jordan, Jones, & Goulding Inc.
KCI Technologies Inc.
Kelly & Weaver, P.C.
Kennedy/Jenks Consultants
Komline Sanderson Engineering Corporation
Lawler, Matusky & Skelly Engineers, LLP
LTI Inc.
Lombardo Associates Inc.
Malcolm Pirnie Inc.
McKim & Creed
MEC Analytical Systems Inc.
Metcalf & Eddy Inc.
MWH
New England Organics
O'Brien & Gere Engineers Inc.
Odor & Corrosion Technology Consultants Inc.
Oswald Green, LLC
PA Government Services Inc.
Parametrix Inc.
Parsons
Post, Buckley, Schuh & Jernigan
R&D Engineering/Conestoga Rover & Associates
RMC Inc.
R.M. Towill Corporation
Ross & Associates Ltd.
Royce Technologies
SAIC Maritime Technical Group
Short Elliott Hendrickson Inc.
Stantec Consulting Inc.
Stearns & Wheler, LLC
Stone Environmental Inc.
Stormwater Management Inc.
Stratus Consulting Inc.
Synagro Technologies Inc.
Tetra Tech Inc.
Trojan Technologies Inc.
URS Corporation
USFilter
Veola Water North America
Wade-Trim Inc.
Westin Engineering Inc.
Weston Solutions Inc.
Woodard & Curran
WRc/D&B, LLC
WWETCO, LLC
Zenon Environmental Inc.
Zoeller Pump Company

INDUSTRY

American Electric Power
ChevronTexaco Energy Research & Technology Company
The Coca-Cola Company
Dow Chemical Company
DuPont Company
Eastman Chemical company
Eastman Kodak Company
Eli Lilly & Company
Merck & Company Inc.
ONDEO Services
Procter & Gamble Company
PSEG Services Corporation
RWE Thames Water Plc
Severn Trent Services Inc.
United Water Services LLC

Note: List as of 2/1/05

WERF Product Order Form

As a benefit of joining the Water Environment Research Foundation, subscribers are entitled to receive one complimentary copy of all final reports and other products.Additional copies are available at cost (usually $10).To order your complimentary copy of a report, please write "free" in the unit price column.WERF keeps track of all orders.If the charge differs from what is shown here, we will call to confirm the total before processing.

Name ____________________ Title ____________________

Organization ____________________

Address ____________________

City __________ State __________ Zip Code __________ Country __________

Phone __________ Fax __________ Email __________

Stock #	Product	Quantity	Unit Price	Total
			Postage & Handling	
			VA Residents Add 4.5% Sales Tax	
			Canadian Residents Add 7% GST	
			TOTAL	

Method of Payment: (All orders must be prepaid.)

☐ Check or Money Order Enclosed

☐ Visa ☐ Mastercard ☐ American Express

Account No. __________ Exp. Date __________

Signature __________

Shipping & Handling:

Amount of Order	United States	Canada & Mexico	All Others
Up to but not more than:	Add:	Add:	Add:
$20.00	$5.00*	$8.00	50% of amount
30.00	5.50	8.00	40% of amount
40.00	6.00	8.00	
50.00	6.50	14.00	
60.00	7.00	14.00	
80.00	8.00	14.00	
100.00	10.00	21.00	
150.00	12.50	28.00	
200.00	15.00	35.00	
More than $200.00	Add 20% of order	Add 20% of order	

*minimum amount for all orders

Note: Please make checks payable to the Water Environment Research Foundation.

To Order (Subscribers Only):

Log on to www.werf.org and click on the "Product Catalog."

Phone: (703) 684-2470
Fax: (703) 299-0742.

WERF
Attn: Subscriber Services
635 Slaters Lane
Alexandria, VA 22314-1177

To Order (Non-Subscribers):

Non-subscribers may be able to order WERF publications either through WEF (www.wef.org) or IWAP (www.iwapublishing.com).Visit WERF's website at www.werf.org for details.

IWA Publishing's authorised EU representative for General Product Safety Regulations is Diane D'Arras, 15 rue Duret, 75116 Paris, France, e-mail: safety@iwap.co.uk.

Printed and bound by CPI Group (UK) Ltd, Croydon, CR0 4YY
12/05/2026
02108893-0010